SELF ASSEMBLY

The Science of Things That Put Themselves Together

John A. Pelesko

Chapman & Hall/CRC
Taylor & Francis Group
Boca Raton London New York

Chapman & Hall/CRC is an imprint of the
Taylor & Francis Group, an informa business

Chapman & Hall/CRC
Taylor & Francis Group
6000 Broken Sound Parkway NW, Suite 300
Boca Raton, FL 33487-2742

© 2007 by Taylor & Francis Group, LLC
Chapman & Hall/CRC is an imprint of Taylor & Francis Group, an Informa business

No claim to original U.S. Government works
Printed in the United States of America on acid-free paper
10 9 8 7 6 5 4 3 2 1

International Standard Book Number-10: 1-58488-687-0 (Softcover)
International Standard Book Number-13: 978-1-58488-687-7 (Softcover)

Visit the Taylor & Francis Web site at
http://www.taylorandfrancis.com

and the CRC Press Web site at
http://www.crcpress.com

Contents

Preface

Nanotechnology has been hailed as the "21st Century's great leap forward in scientific knowledge." It's claimed that nanotechnology "is destined to become the core technology underlying all of 21st century medicine" and that it will "cure cancer and replace fossil fuels" and yet that these advances will only "seem a minor part of the whole."[1] I don't know where nanotechnology will take us and I don't know whether the claims of its most avid proponents will prove true, but I do know that the science of *self-assembly* offers the most promising route to true molecular nanotechnology available today. This book is about self-assembly.

As an enabling technique for nanotechnology, self-assembly replaces *top down* fabrication with the possibility of *bottom up* fabrication. It's the difference between building nanoscale structures molecule by molecule using the equivalent of nano-chopsticks, and letting molecules do what they do best, self-assemble themselves into useful structures. But, to fully utilize this new technique, we must understand what it is that nature does when she builds objects by self-assembly. In recent years, a host of scientists and engineers, from every imaginable discipline, have been working to figure this out. And, more importantly, they've been working to put nature's principles to use in the laboratory and eventually in the factory. While much remains to be done, much has been accomplished. It's these accomplishments that we'll look at in this book.

This book is divided into three parts. This structure reflects the natural progress of the science of self-assembly. We begin in *Part I: The Natural World*, by examining just what it is that nature does and what we can say about how she does it. With this inspiration, we move on to *Part II: Engineered Systems*, and examine the many different ways in which scientists are learning to exploit nature's techniques. Finally, in *Part III: The Future*, we examine the developing theoretical underpinnings of self-assembly and the latest varied advances in the field. It is through this pairing, theory and experiment, that the science of self-assembly is moving forward.

But, self-assembly is not just an enabling technique for nanotechnology. The science of self-assembly is also a way of understanding the natural world, understanding biology, understanding physical phenomena, and perhaps, ul-

---

[1]These quotes are from *Nanotechnology* by Wilson et al. the article *Nanomedicine* by Ralph Freitas, and the article *Why should you care about molecular nanotechnology?* by K. Eric Drexler, respectively.

timately, understanding the origin of life itself. Consequently, this book often goes beyond nanotechnology - we examine systems at all length scales. The greatest practical application of these efforts may lie in nanoscience, but the most important implications of the ideas explored here may lay elsewhere.

Self Assembly: The Science of Things that Put Themselves Together, is an introduction to this exciting field. It is intended for use by working scientists and engineers in every discipline, as well as students studying science, engineering, or mathematics who wish to understand the science of self-assembly as both the great enabling technology for nanotechnology and as a viewpoint for understanding many features of the natural world.

Prerequisites

As with nanotechnology in general, self-assembly does not fit into any convenient academic box. Research in self-assembly is being conducted by individuals in every imaginable discipline. To grasp the details of every study that has been done one would need an extensive background in chemistry, physics, biology, computer science, mathematics, and engineering. Of course, for most of us, that's just not possible. So, as much as possible, this book has been written to be self-contained. Concepts from different disciplines are introduced as needed and explained in sufficient detail to allow the reader to grasp the main ideas of the text. The Related Reading section at the end of each chapter gives suggestions for following up on topics or filling in needed background. References to all of the original journal articles discussed in the text are given at the appropriate point. The reader may find the glossary at the end of the text useful; there is also an appendix that can be used to quickly introduce the reader to the calculus of variations.

Chapter Interdependence

This book is meant to be read sequentially. Ideas build from chapter to chapter, examples become increasingly complex, and exercises rely on information presented in earlier chapters. Nonetheless, instructors may find that the background of their students allows them to skip certain chapters or integrate the material of later chapters with the material of former chapters. For a class of students with little mathematical background, Chapter 9 may be skipped entirely. The basic modelling ideas introduced in Chapter 9 are introduced elsewhere, but with much less detail. For an advanced class, the instructor may wish to include material from Chapter 9 earlier in the discussion. For example, the graph grammar model of Chapter 9 can be discussed immediately following Chapter 7. Similarly, the conformational switch model of Chapter 9 can be brought in as early as Chapter 3.

Profiles

In every chapter of this text you'll find a one-page profile of a notable individual who has made important contributions to the field of self-assembly. Unfortunately, there are only ten chapters and hence I could only include ten individuals. There are many more noteworthy people working in this area than these ten. Some may find my choices here unusual. With the exception of Richard P. Feynman, profiled in Chapter 1, all of the profiles are of active researchers. Many of them are young, several (gasp!) are not even yet full professors. Or, professors of any kind. Yet, each individual profiled has something important to say about the science of self-assembly. I hope you'll enjoy meeting them.

Try it Yourself

Self-assembly lends itself to hands-on activities that can greatly enhance your understanding and feel for the problems and challenges of this field. Scattered throughout the text you'll find *Try it Yourself* exercises. Each of these requires a minimum of experimental expertise, can usually be built with everyday or easily acquirable materials, and serves to demonstrate at least one key principle discussed in the text. With the aid of many undergraduate students working in our lab at the University of Delaware, I've carried out all of the experiments described here. If you have difficulty following the instructions or getting your system to work as described please feel free to contact me through the web page for this text. The web page also contains video of several of the experiments and links to other pages illustrating these experiments.

Related Reading

At the end of each chapter you'll find a section entitled *Related Reading*. These sections contain pointers to books, journal articles, and web pages that relate to and expand upon the material in the given chapter. I've resisted the temptation, succumbed to by far too many authors, to simply include lists of "classic" works in a given area. Rather, I've followed a strict policy of only recommending books, articles, and web pages that I've read myself, found accessible, and found useful.

The World Wide Web

In only a fraction of my lifetime the world wide web has evolved from a novelty to an essential feature of any text such as this. To accompany this book, I've developed a web page, www.pelesko.com, that contains links to the numerous web sites mentioned in the text as well as other resources and information about self-assembly. As you read the text, you'll encounter several systems that can only truly be appreciated by watching them in action. Links to video of these systems may be found on the web page for this book. When you reach these points in the text, I encourage you to go to the web page and view the video of the relevant system.

A Note on Terminology

A majority of authors seem to use a hyphen when referring to "self-assembly" or "self-assembled" systems. A minority omit the hyphen. With the exception of the title, I've deferred to the majority.

Acknowledgments

I am indebted to many people whose advice, support, and help during the writing of this book have been invaluable. Thanks to my editor at CRC Press, Sunil Nair, and all of the staff at CRC, especially Helena Redshaw. Thanks to Rob Ghrist for many conversations that forced me to think broadly about the meaning of self-assembly. Thanks to the many students who have been indirectly or directly involved in my exploration of self-assembly. Especially, the students of Math 824 in Spring 2004, and the many undergraduate and graduate students who have worked in the MEC Lab. Among these, Katy Sharpe, Janine Janoski, Lauren Rossi, Dan Cargill, Tom Fleetman, Derick Moulton, and Regan Beckham deserve special mention.

I owe a special debt of thanks to all of those who agreed to be profiled in this book. Especially, Erik Klavins, Eric Winfree, Paul Rothemund, Saul Griffith, Ned Seeman, and John Reif for their gift of time.

Their are many wonderful people I have interacted with over the years who have shaped my career. I would especially like to acknowledge Greg Kriegsmann and Daljit Ahluwalia for the environment they created at the New Jersey Institute of Technology. Thanks to my colleagues at the University of Delaware, especially Toby Driscoll and Lou Rossi for making UD such a fun place to work. I need to thank my high school English teacher, Julius Gottilla, who agreed to finally read something I wrote as long as his name appeared in the acknowledgements. We'll see. I'd like to thank Julia Lee Pelesko for "lending me a hand" with Chapter 3. I'd like to thank John R. Pelesko for letting me use his bath time to explore bubble rafts and inspiring the start to Chapter 2. But mostly, I'd like to thank Julia and John-John for the time I spent writing that I could have otherwise spent with them.

This book is dedicated to the memory of John Pelesko, Sr., 1916-2005, hobo, soldier, cowboy, trucker, fisherman, father, grandfather, and master storyteller.

John A. Pelesko
Newark, Delaware
January 2007

List of Tables

List of Figures

Chapter 1

Introduction

Though men now possess the power to dominate and exploit every corner of the natural world, nothing in that fact implies that they have the right or the need to do so.

Edward Abbey

1.1 Self-Assembly

This is a book about an idea. An observation, really. Simply stated - *No one put you together.* Or the trees outside my window, or the groundhog under my shed, or the simplest bacteria, or the largest whale, or the salt on my popcorn, or the soap bubbles I blow for my children. Somehow, remarkably, all of these things, some alive, some not, *put themselves together.* We call this process, this idea, *self-assembly.* Today, this simple observation has become the basis for one of the most exciting research directions in science, and more modestly, the subject of this book - *Self Assembly: The Science of Things that Put Themselves Together.*

Understanding self-assembly requires the efforts of researchers from almost every imaginable discipline. Biologists are busy unravelling nature's secrets, yielding a deeper understanding of how she effortlessly produces intricate structures from simple building blocks. Chemists are coaxing molecules to form into ever larger and more complex systems. Engineers are developing new manufacturing methods, pushing the boundaries of engineered systems to the nanoscale. Computer scientists are learning to compute with DNA, while mathematicians are developing models to help solve the difficult design problems we encounter as we learn to harness the power of self-assembly. The study of self-assembly is truly a multi-disciplinary endeavor. If you want to understand what self-assembly is, if you are excited by the simple idea that no one put you together, and if you are prepared to examine this subject from a wide variety of perspectives, then this book is for you.

But, self-assembly is a slippery concept. Patterns and structures abound in nature. What is self-assembled? And, just as importantly, what is not? Consider the picture of the barred spiral galaxy NGC 1365 shown in Figure

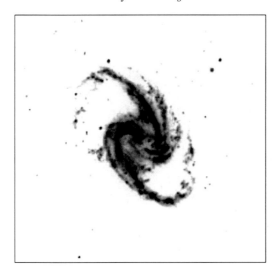

FIGURE 1.1:   Barred Spiral Galaxy NGC 1365. Colors have been reversed to highlight the spiral structure. Credit: NASA Jet Propulsion Laboratory.

1.1. The spiral structure is evident, and we know that this structure is made up of billions of individual particles, stars. Under the influence of gravitational forces, these billions of stars have organized themselves into the structure we see in Figure 1.1. Is this self-assembly? Or, consider the Von Karman vortices over Alaska shown in Color Plate 11.2. Again, a pattern is evident. We know that these structures were formed as the atmosphere interacted with Alaska's Aleutian Islands. Is this self-assembly? On a smaller scale, consider the fluoride crystal of Color Plate 11.6, or the bismuth crystal of Color Plate 11.4. Should crystal growth be considered self-assembly? Or, we can look to living systems. Did the pattern of spots on the Asian multicolored lady beetle, Color Plate 11.5, arise by some form of self-assembly? Was the structure of the single celled eukaryote Oxytricha Trifallax, Figure 1.2, built by some self-assembly process?

Clearly, in nature, the potential range of phenomena that could be called "self-assembly" is enormous. How about in man-made systems? Unfortunately, researchers in almost every discipline use the term "self-assembly," and even more unfortunately, they often use it to describe very different ideas. We need to focus. We need a guiding principle, a concrete definition, and a goal.

In this book, our guiding principle is this - we believe there is a growing body of researchers, from a wide variety of disciplines, investigating something new and exciting called "self-assembly," and ultimately, although their approaches may differ, they are talking about the same thing. It's a bit like the fable of the six blind men and the elephant[1]. Chemists are feeling the tail and declaring "supermolecular chemistry!" Biologists are squeezing the nose and

FIGURE 1.2:   The single celled eukaryote Oxytricha Trifallax. Credit: National Institutes of Health.

exclaiming "protein folding!" Meanwhile mathematicians are flapping the ears and muttering about Wang Tiles and Universal Turing Machines. But, in the end, it's all elephant. In this book, the belief that a new cross-cutting discipline is emerging and that this discipline should be called "self-assembly" will serve as our guide.

Our goal is more concrete; we want to understand how nature self-assembles structures, we want to understand her principles and techniques, and, we want to learn how to use self-assembly to build engineered systems. The structure of *Self Assembly: The Science of Things that Put Themselves Together* reflects this goal. In *Part I: The Natural World*, we'll take a closer look at natural self-assembling systems. We address the question - What does nature do and how does she do it? We'll begin, in Chapter 2, with inorganic systems and examine in detail topics such as crystal growth, soap films, and micelles. In Chapter 3, we'll look at organic systems and see how nature builds viruses, proteins, and ribosomes. We'll introduce the protein folding problem and see how nature uses energy minimization to produce a remarkable range of biologically functional molecules. In the concluding chapter of Part I, Chapter 4, we'll examine what we've learned and abstract the principles nature uses to self-assemble structures.

In *Part II: Engineered Systems*, we ask the question - What can we build now and how do we do it? We'll see how nature's principles are being applied by physicists, chemists, biologists, and engineers as they induce cubes to self-assemble from DNA or electronic circuits to self-assemble from millimeter-scale polyhedra. In the first chapter of Part II, Chapter 5, we examine the simplest engineered self-assembling systems. What we learn from these simple systems allows us to understand the more complicated systems of Chapters 6,

7, and 8. In Chapter 6, we'll see how capillary forces, magnetic forces, and the principle of energy minimization are used to design functional self-assembling systems. In Chapter 7, we focus on dynamic systems, i.e., those that self-assemble and maintain organization only while dissipating energy. Finally, in Chapter 8, we address the myriad ways in which DNA is being exploited in the design of self-assembled systems.

In the final part of this book, *Part III: The Future*, we ask the question - How do we realize the full promise of self-assembly? We'll revisit the challenges in understanding how nature self-assembles systems and the challenges in designing engineered self-assembling systems that we encountered in the first two parts of this book as we survey the various theoretical approaches and the latest experimental approaches to overcoming these challenges. Hopefully, by this point, you'll have a clear understanding of what self-assembly is, and be fully prepared to delve into the primary literature in the field.

Now, we have our principle and we have our goal, but we still need our definition. Just as Aristotle grappled with parsing the difference between the animate and the inanimate, or between man and animal, scientists today have grappled with how best to define the burgeoning field of self-assembly. While no present definition of self-assembly approaches the elegance of Aristotle's definition of man as a "rational animal," they are all worth a look.

In a relatively early paper on self-assembly Hosokawa et al. [62] refer to nature in an attempt to define self-assembly:

> Viruses and bacterial flagella are constructed automatically out of protein subunits. This phenomenon is called self-assembly, which is a powerful technique applicable to microfabrication.

They go on and enumerate the conditions necessary for a system to self-assemble:

> To achieve self-assembly, the following conditions must be met: generating bonding forces, bonding selectively, and moving the parts randomly so that they come together by chance.

A later simpler definition is offered by Campbell et al. [25]:

> Spontaneous assembly, often called "self-assembly," refers to aggregation of particles into an organized structure without external assistance.

Equally simple definitions have been offered by several others. Aggarwal et al. states [3]:

> Self-assembly is the ubiquitous process by which objects autonomously assemble into complexes.

John H. Reif [102] also offers a straightforward definition:

Self-assembly is a process in which small objects autonomously associate with each other to form larger complexes.

The group led by George M. Whitesides of Harvard University has offered several definitions. In a 2002 review paper [139] they ask the question "Is anything not self-assembly?" and offer this definition in an attempt to distinguish between self-assembly and formation:

...we limit the term to processes that involve pre-existing components, are reversible, and can be controlled by the proper design of the components.

Finally, in a 2006 publication [41] the group led by Bartosz A. Grzybowski at Northwestern University defined self-assembly as:

...we limit SA to the spontaneous formation of organized structures from many discrete components that interact with one another directly and/or indirectly through their environment. In addition, the assembling components may also be subject to various global (confining) potentials such as externally imposed electromagnetic fields or chemical gradients.

Freely borrowing from all of the above, in this book we define self-assembly as follows. **Self-assembly** *refers to the spontaneous formation of organized structures through a stochastic process that involves pre-existing components, is reversible, and can be controlled by proper design of the components, the environment, and the driving force.* As we make our way through this text, we'll see the necessity of each of the elements of this definition. The word "organized" will allow us to distinguish between self-assembly and aggregation processes. The emphasis on "pre-existing components" will allow us to distinguish between self-assembly and pattern formation while the words "stochastic," "design," "environment," and "driving force" push us towards an understanding of those features of self-assembly necessary to reach our goal of designing engineered self-assembling systems.

Finally, within the class of phenomena that we call self-assembly, it is useful to emphasize three particular subclasses. *By* **static self-assembly** *we mean that subclass of self-assembly processes that leads to structures in either local or global equilibrium. By* **dynamic self-assembly** *we mean that subclass of self-assembly processes that leads to stable non-equilibrium structures. That* is, these structures exist only so long as the system is dissipating energy. *By* **programmed or programmable self-assembly** *we mean that subclass of self-assembly processes where the particles of the system carry information about the final desired structure or its function.*

Profile - Richard P. Feynman

With the title of his 1992 biography of Richard P. Feynman, author James Gleick, succinctly described this Noble-Prize winning physicist and father of nanotechnology: *Genius*. Feynman was born in New York City on May 11th, 1918. He grew up in Far Rockaway, where even as a youngster he established a reputation for his unbridled curiosity, his sense or humor, and a talent for mathematics. As an undergraduate, Feynman attended the Massachusetts Institute of Technology. He received his Ph.D. from Princeton University in 1942. He subsequently held appointments at Cornell University and the California Institute of Technology. During World War II, Feynman played a key role in the Manhattan Project, helping to develop the first atomic bomb. In 1965, Feynman won the Noble Prize in Physics for his work on quantum electrodynamics. A far-ranging thinker, Feynman is remembered for much more than just his contributions to quantum theory. He made fundamental discoveries in superfluidity, the theory of quarks, and the theory of the weak nuclear force. He served on the President's commission investigating the space-shuttle *Challenger* disaster. His elegant demonstration of the reason for the shuttle failure, dipping an O-ring in a glass of ice-water, illustrated his ability as an educator and his talent for making difficult concepts clear. The Foresight Nanotech Institute created two prizes named after Richard P. Feynman, acknowledging his role in launching the field of nanotechnology. The Foresight Institute Feynman prize is awarded annually to one theoretician and one experimentalist who has advanced the achievement of Feynman's vision for nanotechnology. The as of yet unclaimed Feynman Grand Prize will be awarded to the first team who designs and builds a nanoscale robotic arm and a nanoscale computer.

No brief biography or profile can truly capture the essence of Richard P. Feynman or the reason for the extent of his influence on both the scientific and lay communities. Feynman's advice to the Caltech graduating class of 1974 may do a better job:

> *The first principle is that you must not fool yourself – and you are the easiest person to fool.*

Further Reading There have been numerous books written about Richard P. Feynman. The best biography is the one by Gleick [48]. However, we are also fortunate to have access to Feynman through his own writing and recorded lectures. He recounts many of his adventures as a youth, at MIT, and as a Ph.D. student in his bestselling 1985 autobiography, *Surely You're Joking, Mr. Feynman!* This was followed by the entertaining *What Do You Care What Other People Think?* in 1988. Many of his lectures were captured on audio tape and have been released on compact disc. Several are available as free streaming-video on the internet. Links may be found on the web page for this book. Finally, any serious student of physics must own the three volume set, *The Feynman Lectures on Physics* [42].

1.2   Why Now?

Now that we have a working definition of self-assembly, the question still remains: Why the sudden explosion of interest in this field? That self-assembly has thoroughly captured the interest of the scientific community was made clear with the publication of the 125th anniversary issue of *Science*. To celebrate this anniversary, the editors identified twenty-five big questions and one-hundred little questions likely to shape the course of scientific research for the next one-hundred and twenty-five years. Listed among the big questions, right alongside "What is the universe made of?" "Can the laws of physics be unified?" and "Are we alone in the universe?" we find "How far can we push chemical self-assembly?" How did self-assembly, a field yet barely defined, rise to prominence so quickly? This sudden ascent may be ascribed to a confluence of developments in science and engineering.

First among these is the advent of *nanotechnology*. Nanotechnology deals with the very small, with the construction of structures characterized by length scales of less than 100nm. To get a sense of this scale, let's contrast a nanoscale machine that we'll study in Chapter 3, the ribosome, with some everyday small objects. A typical ribosome measures about 25nm in diameter. Keep in mind that the ribosome is a *fully-functional machine*. Contrast the size of the ribosome with that of a typical grain of sand. On average, sand is about $500\mu m$ in diameter. That's $500,000$nm, $20,000$ times the size of the ribosome. Human hair brings us a little closer. With an average diameter of around $50\mu m$ or $50,000$nm, that's only 2000 times the size of the ribosome. Nanotechnologists are attempting to replicate nature's ability to make useful machines, such as the ribosome, on the nanometer scale.

The possibility that humans could build nanoscale machines was first recognized by Richard P. Feynman and discussed in a famous lecture at the 1959 annual meeting of the American Physical Society. In a passage on biology, Feynman captured much of what excites self-assembly researchers today:

> The biological example of writing information on a small scale has inspired me to think of something that should be possible. Biology is not simply writing information; it is *doing something* about it. A biological system can be exceedingly small. Many of the cells are very tiny, but they are very active; they manufacture various substances; they walk around; they wiggle; and they do all kinds of marvellous things – all on a very small scale. Also, they store information. Consider the possibility that we too can make a thing very small which does what we want – that we can manufacture an object that maneuvers at that level!

While he did not anticipate or discuss the notion of self-assembly, Feynman did recognize that the problem of actually making nanoscale systems was a

TABLE 1.1:   Landmarks in the History of Nanotechnology Reproduced
with permission from *Modeling MEMS and NEMS*, Pelesko and Bernstein [99].

1940s	Radar drives the development of pure semiconductors.
1959	Richard P. Feynman's famous "There's plenty of room at the bottom" lecture.
1960	Planar batch-fabrication process invented.
1964	H.C. Nathanson and team at Westinghouse produce the resonant gate transistor, the first batch-fabricated MEMS device.
1970	The microprocessor is invented, driving the demand for integrated circuits ever higher.
1979	The first micromachined accelerometer is developed at Stanford University.
1981	K. Eric Drexler's article, *Molecular Engineering: An Approach to the Development of General Capabilities for Molecular Manipulation*, is published in the Proceedings of the National Academy of Sciences. This is arguably the first journal article on molecular nanotechnology to appear.
1982	The scanning tunneling microscope is invented.
1984	The polysilicon surface micromachining process is developed at the University of California, Berkeley. MEMS and integrated circuits can be fabricated together for the first time.
1985	The "Buckyball" is discovered.
1986	The atomic force microscope is invented.
1991	The carbon nanotube is discovered.
1996	Richard Smalley develops a technique for producing carbon nanotubes of uniform diameter.
2000s	The number of MEMS devices and applications continually increases. National attention is focused on funding nanotechnology research and education.

difficult one. In some sense, in the fifty years since Feynman's lecture, we've made rapid progress in tackling this problem. Today, we have batch-fabricated microprocessors containing nanoscale transistors in our cell phones. We have new tools such as the scanning tunnelling microscope (STM) and the atomic force microscope (AFM) that allow us to examine and manipulate matter on the nanoscale. And, we have new nanostructured materials such as the carbon nanotube that promise to revolutionize materials science. But, researchers in nanotechnology have come to realize that for all of the progress we've made, we still rely on "top-down" construction methods. When Eigler and Schweizer [37] famously wrote the letters "IBM" on a layer of nickel using individual xenon atoms, a true tour-de-force in nanoscale engineering, they still used a fundamentally primitive and decidedly unbiological technology. In essence, Eigler and Schweizer used their STM as a pair of very small tweezers and wrote "IBM" much as I might spell out my name on my desk with grains of salt. Researchers in nanotechnology have come to realize that if we are to truly realize Feynman's vision, we not only have to learn to build small, we have to learn to build small in the way that nature builds small. We need to coax our systems to self-assemble.

In parallel with progress in nanotechnology fuelling the need for understanding self-assembly, key developments in mathematics, computer science,

TABLE 1.2: Landmarks in Self-Assembly

1930s	Alan Turing develops the theory of universal computation.
1950s	John von Neumann develops theory of automata replication.
1953	James D. Watson and Francis Crick discover the structure of DNA.
1955	H. Fraenkel-Conrat and R.C. Williams self-assemble the tobacco mosaic virus in a test tube.
1957	Penrose and Penrose construct a simple self-replicating system.
1961	Hao Wang develops "Wang Tiles" demonstrating the equivalence of tiling problems and computation.
1991	Nadrian C. Seeman and Junghuei Chen self-assemble a cube from DNA.
1994	Leonard Adleman launches the field of DNA computation by using DNA to solve a Hamiltonian path problem.
1996	Kazuo Hosokawa's group demonstrates microscale self-assembly using surface tension.
2000	George M. Whitesides's group self-assembles electrical networks from millimeter scale polyhedra.
2004	William Shih adapts the methods of Seeman to self-assemble a DNA octahedron.
2004	Eric Winfree and Paul Rothemund self-assemble a Sierpinski triangle from DNA demonstrating that self-assembly may be used for computation.
2000s	Self-assembly research explodes drawing the interest of researchers from every imaginable field.

biology, and chemistry have brought us to the point where it is becoming possible to begin to understand and utilize self-assembly. Curiously, some of the key developments took place almost contemporaneously with Feynman's lecture. In the 1930's, the British mathematician Alan Turing, developed the theory of universal computation. Long before the arrival of the personal computer, Turing had already liberated computation from the silicon chips with which we so closely associate computation today. With his "Universal Turing Machine," Turing taught us that computation could be thought of abstractly and that all sufficiently complex computers are essentially equivalent. His insight paved the way for Adleman's invention of DNA computing in 1994 and Winfree's 2004 demonstration of computing via self-assembly. In the 1950's the mathematician John von Neumann extended Turing's efforts and developed the theory of automata replication. Von Neumann's work created a framework for future efforts in the development of self-replication, artifical life, and self-assembly. Another key landmark in the history of self-assembly was the discovery of the structure of DNA by Watson and Crick in 1953. DNA, the central molecule of biology, is also of central importance in the study of self-assembly. In addition to being used for computation, it was shown by Nadrian C. Seeman in 1991 that DNA could be induced to self-assemble into mechanical structures. The self-assembled DNA cube of Seeman and Chen has already been improved upon by researchers such as William Shih who has coaxed DNA into self-assembling into octahedra and other complex shapes. Shortly after the discovery of the structure of DNA, Fraenkel-Conrat and

Williams showed that biological systems could be induced to self-assemble in a test tube. There work with the tobacco mosaic virus allowed us to begin to understand how nature uses self-assembly in biology. Another early key development in self-assembly, that occurred shortly after Feynman's lecture, was the invention of "Wang Tiles" by Hao Wang. Wang showed the equivalence of tiling problems and computation, thereby extending Turing's work and providing the second key ingredient for Winfree's demonstration of computation by self-assembly. On the macroscale, other researchers have made fundamental contributions to our understanding and practical implementation of self-assembly. Notable among these is the invention of a simple self-replicating machine by Penrose and Penrose in 1957, the use of surface tension to self-assemble 2-d structures by Hosokawa in 1996, and the practical implementation of surface tension driven self-assembly by Whitesides in 2000. Today, self-assembly is drawing the interest and efforts of researchers from every imaginable discipline. While we are still a long way from duplicating the elegance of nature, we're closer than ever, and getting closer every day.

1.3 Chapter Highlights

- Self-assembly is a multi-disciplinary endeavor. If you want to understand this field and examine self-assembly from a wide variety of perspectives, this book is for you.

- **Self-assembly** refers to the spontaneous formation of organized structures through a stochastic process that involves pre-existing components, is reversible, and can be controlled by proper design of the components, the environment, and the driving force.

- **Static self-assembly** refers to that subclass of self-assembly processes that leads to structures in local or global equilibrium.

- **Dynamic self-assembly** refers to that subclass of self-assembly processes that leads to stable non-equilibrium structures. These structures persist only so long as the system is dissipating energy.

- **Programmed or programmable self-assembly** refers to that subclass of self-assembly processes where the particles of the system carry information about the final desired structure or its function.

- Numerous examples of self-assembling systems may be found in the natural world. These include both organic and inorganic systems. Part I of this text deals with natural self-assembling systems.

- Engineers are learning to exploit the power of self-assembly in the production of manmade systems. Part II of this text focuses on engineered self-assembling systems.

- The gap between what nature can produce and what man can produce is large. Part III of this text focuses on approaches to closing this gap.

1.4 Exercises

Section 1.1

1. There is a large body of research on "pattern formation." At the heart of much of this work lies the Turing instability. Investigate the Turing instability and explain why these pattern formation mechanisms are not considered self-assembly as defined in this book.

2. The cartoon by Saul T. Griffith shown in Color Plate 11.1 gives a simple example of a self-assembling system. Compare this system to our definition of self-assembly, and explain how each term in our definition has a counterpart in this system. Is this an example of static or dynamic self-assembly?

3. Nature provides us with many potential examples of self-assembly. Consider the hailstone of Figure 1.3. How does this hailstone grow? Is this self-assembly?

Section 1.2

4. Read Feynman's "There's plenty of room at the bottom" lecture. The reference may be found in the Related Reading section of this chapter. What methods does Feynman propose for constructing objects at the nanoscale. How do they compare to self-assembly?

1.5 Related Reading

Richard P. Feynman's classic article is timeless and well-worth reading.

R.P. Feynman, *There's Plenty of Room at the Bottom*, J. Microelectromechanical Sys., 1 (1992), pp. 60-66.

FIGURE 1.3: Grapefruit sized hailstone. Note the smaller hailstones comprising the aggregate. Credit: NOAA Photo Library, NOAA Central Library; OAR/ERL/National Severe Storms Laboratory (NSSL) .

Two entertaining popular treatments of nanotechnology are the books by Regis and Gross. The book by Regis contains an interesting account of Richard Feynman's lecture on nanotechnology and the effect of his talk on the scientific community.

E. Regis, *Nano: The Emerging Science of Nanotechnology*, Little, Brown and Company, 1995.

M. Gross, *Travels to the Nanoworld*, Plenum Trade, 1995.

An excellent collection of articles on nanotechnology that includes a brief description of self-assembly was edited by Wilson et al.

N. Wilson et al. , *Nanotechnology: Basic science and emerging technologies*, Chapman and Hall/CRC, 2002.

The Whitesides' group has produced several must-read review papers on self-assembly. The most accessible appeared in Science.

G.M. Whitesides and B. Grzybowski, *Self-Assembly at All Scales*, Science, 295 (2002), pp. 2418-2421.

1.6 Notes

1. A nice retelling of the original fable may be found in *The Moral Compass* by William J. Bennett. The fable was also adapted by John Godfrey Saxe and recast as a poem.

Part I

The Natural World

Chapter 2

Inorganic Systems

All nature... as it exists by itself, is founded on two things: there are bodies and there is void in which these bodies are placed and through which they move about.

Lucretius, *Nature of Things*

2.1   Introduction

It is not difficult to find patterns in nature. In fact, quite the opposite is true. But, not all natural patterns are the result of a self-assembly process. In this chapter we examine four naturally occurring inorganic systems that are the result of some form of self-assembly. Our goal is to begin to understand the principles nature uses when she induces objects to self-assemble.

We begin in Section 2.2 with *bubble rafts.* This very simple system is easily observed in nature or in the laboratory and often used in the classroom to illustrate the bonding and packing of atoms that occurs in crystallization. We describe the bubble raft and take a first look at the capillary forces driving the formation of the raft. We'll revisit capillary forces in Part II when we examine engineered self-assembling systems. We also encounter the notion of *packing.* While understanding capillary forces allows us to understand why bubbles coalesce, understanding packing allows us to understand the global arrangement of bubbles in a bubble raft. This theme of packing as a minimal energy configuration will reoccur throughout the text. Next, in Section 2.3 we turn to *crystallization.* With the bubble-raft model in hand, we describe the basics of crystallization and examine the order exhibited by crystals on both the atomic and macroscopic levels. We introduce the notion of *diffusion limited aggregation* (DLA). This serves as our first computational model of a self-assembly process. In Section 2.4 we turn our attention to *polymerization.* Polymers are a class of materials of tremendous technological importance and in fact could be appropriately treated in this chapter, in Chapter 3 when we consider organic systems, or in Part II when we consider engineered systems. That is, there are naturally occurring inorganic polymers, naturally occurring organic polymers, and engineered polymers. We introduce polymers here in order to be able to present a second model of a self-assembling system. We examine a simple reaction-kinetics type model based on the Law of Mass Action, solve this model, and consider the implications of this model for self-assembly. This type of model will arise again in Parts II and III of this text. Finally, in Section 2.5 we examine the remarkable self-assembling *micelle*. Micelles are structures formed in solution from amphiphilic molecules. These molecules consist of a water loving head group and an oil loving tail. When placed in water in the proper concentration, these molecules spontaneously assemble into spherical structures, tube-like structures, and eventually "living polymer" superstructures. When placed in oil, analogous reverse-micelles are formed. Both the process of micelle formation and the micelles themselves are the subject of much active research. Their similarity to biological membranes promises to make micelles of great use in applications as well as a great help in understanding the process of self-assembly.

2.2   Bubble Rafts

It is darn near impossible to live on the planet Earth and never encounter a bubble. We see bubbles in the sink, we see bubbles in the shower, we see bubbles on the stove top; small children blow bubbles with their saliva, older children blow bubbles from soap and water, and still older children blow bubbles from chewing gum. At one time or another every even mildly curious person has surely contemplated the perfection of a soap bubble, the beauty of its swirling colors, and its magical fragility.

Perhaps less often contemplated is the different sort of magic that occurs when multiple bubbles form and interact on the surface of a liquid. It is not that this phenomenon is outside the realm of everyday experience; pour yourself a glass of soda or a beer and you'll encounter many bubbles interacting. But, the process is dynamic, the bubbles are often of very different sizes, and the regularity of the structures formed less apparent to the eye under uncontrolled conditions.

Nonetheless, it is easy to realize that *something* is happening. If we look at soap bubbles in a bathtub, we're likely to see structures like the one shown in Figure 2.1. Most of the bubbles are clustered together. We'd see the lone bubble off to the left approaching the cluster. If we popped one of the larger bubbles, we'd quickly see the vacancy filled as the cluster rearranged itself into some new tightly packed configuration. It was probably observations such as these, together with a desire to build a macroscopic model of crystal structure that led Bragg and Nye [20, 21] to invent the so-called *bubble raft* in the 1940's.[1] Their experimental setup was simple and in the *Try It Yourself* section of this chapter you'll find instructions for making your own bubble

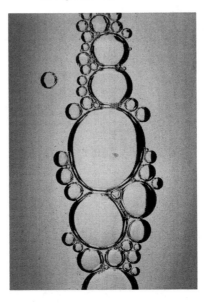

FIGURE 2.1: Bubbles aggregating on the surface of water. Credit: iStock-Photo.com/Steve Gray

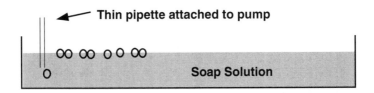

FIGURE 2.2: The basic experimental setup for constructing a bubble raft.

raft. The basic setup is illustrated in Figure 2.2. A thin layer of soap solution is poured into a tray or dish. A pipette is inserted into the fluid layer and attached to a pump. As the pump is activated, bubbles are formed at the end of the pipette, detach themselves from the tip, rise to the surface of the liquid, and assemble into a bubble raft. The key innovation of Bragg and Nye was developing the pump-pipette technique, allowing them to create bubbles of a uniform size. Once a sufficient number of bubbles formed, a structure similar to that in Figure 2.3 was observed. Note the uniform size of the bubbles and the short and long range order present in the raft. Each bubble is surrounded by six neighbors in a hexagonal pattern. This short range structure is repeated throughout the raft, i.e., the long range order. Bragg and Nye proposed this raft as a new model of crystalline structure. They showed how phenomena such as grain boundaries, dislocations, and

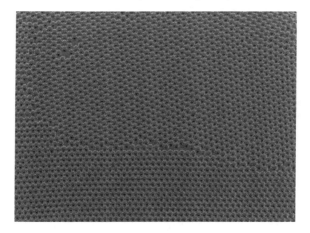

FIGURE 2.3:   A typical bubble raft. Credit: Andrew Lambert Photography /
Photo Researchers, Inc.

recrystallization could be demonstrated using the bubble raft system. Since
that time, the bubble raft has evolved into both a subject of scientific research
and an elegant tool for demonstrating the properties of crystals.

But, here, we are less interested in the bubble raft as a model of a crystal and
more interested in the bubble raft as a self-assembling system. The remarkable
order of Figure 2.3 arose spontaneously. Nothing in the experimental system
tells the bubbles where to go. Yes, it is interesting to investigate the properties
of the fully assembled bubble raft, but we may also ask: How did the bubble
raft self-assemble in the first place? In order to understand the self-assembly
of the bubble raft, we first need to understand the phenomenon of surface
tension.

2.2.1   A Primer on Surface Tension

In his timeless text on soap bubbles [19], C.V. Boys begins by attempting
to convince the reader that the surface of a liquid behaves like an elastic
membrane. He asks the reader to consider a common paint brush. When dry,
the bristles of the brush are separate and randomly arrayed. When wet, the
bristles of the brush cling together. Why? The usual answer is: *because the
bristles are wet.* Boys then asks the reader to consider what the bristles do
when immersed in a glass of water. If you try the experiment for yourself,
you'll find that the bristles do not cling together, but rather they separate and
arrange themselves very much like a dry paintbrush! Why? Are they no longer
wet when placed in a glass of water? Clearly, the real answer as to why they
cling together when wet, but not submerged, is more subtle than originally
thought. In fact, it is not enough for the bristles to be wet, but rather the

brush must be removed from the water so that a liquid surface forms. It is this surface that somehow binds the bristles together. This simple and clear demonstration illustrates the fact that the surface of a liquid behaves differently than the bulk; this is the phenomenon known as *surface tension*.

While surface tension is a macroscopic effect, its origin lies at the microscale. Qualitatively, we can understand the origin of surface tension by considering the environment of molecules in a fluid. Molecules in a fluid exert a variety of forces on one another. These forces are too weak to cause them to bind together, but they effect the behavior of the fluid nonetheless. If a molecule is in the interior of a fluid, far away from any surface, on average, it feels the same force from its neighbors in all directions. Near the surface, things change. A molecule at the surface of a liquid feels less force from the gaseous region above the fluid than it does from its neighbors in the bulk. As a consequence, molecules near the surface feel a net force pulling them back into the bulk. At the macroscale, it appears that the surface *is* behaving like an elastic membrane. The surface exerts a force and attempts to make its area as small as possible.

Mathematically, the concept of surface tension is embodied in the *Laplace-Young Law*[2]

$$\triangle p = \sigma H. \tag{2.1}$$

Here, $\triangle p$ denotes the jump in pressure across a fluid surface, σ is the surface tension, and H is the mean curvature of the surface. The Laplace-Young Law says that the jump in pressure across a fluid surface is proportional to the mean curvature of that surface. The constant of proportionality is an intrinsic property of the fluid and is what we commonly refer to as the surface tension. The $\triangle p$ term has units of force per unit area. The mean curvature H, has units of inverse length. Hence, the surface tension, σ, must have units of force per unit length.

A simple example helps relate the Laplace-Young Law to our everyday experience. Consider a planar circular loop of wire dipped in a soap film. When withdrawn, a flat film will span the loop. This film is a fluid surface. The pressure on both sides of this film is the same, so the Laplace-Young Law reduces to $H = 0$. That is, the mean curvature of the surface must be zero. Of course, a flat surface has zero curvature; Laplace is in agreement with experience.[3]

2.2.2 The Meniscus Effect

An alternative way to look at the Laplace-Young Law or the phenomenon of surface tension is to take an energy or thermodynamic viewpoint. As mentioned above, surface tension causes a liquid to attempt to minimize its surface area. Equivalently, we may say that the liquid attempts to minimize its surface energy. Of course, this must be balanced against other energies present in the system. Overall, the fluid will attempt to minimize its total energy. The behavior of fluids interacting with particles and walls is interesting precisely because of competing energies in the system. This is best illustrated by an example.

If you look closely at the surface of a glass of water, you'll notice that the surface is not flat. This is especially apparent where the liquid comes into contact with the wall. This distortion of the liquid surface, due to the presence of the wall, is known as the *meniscus effect*. Figure 2.4 illustrates this rather dramatically. Using the energy point of view, we can compute

FIGURE 2.4:   A water strider walking on the surface of a pond. Note the meniscus formed near the feet of the bug. Credit: iStockPhoto.com/Sean Lowe.

the shape of this meniscus. We'll consider the idealized system of Figure 2.5. That is, we'll assume we have a semi-infinite bath of fluid in contact with a wall. The wall itself is taken to be infinite in the z-direction so that the shape of the meniscus is only a function of x. The energy, E, of this system is

$$E[u(x)] = \text{Surface Energy} + \text{Gravitational Energy}.$$

The surface energy is proportional to the change in surface area and may be written in terms of the interface shape $u(x)$ as

$$\sigma \left(\int_0^\infty \sqrt{1 + u'^2} dx - A \right). \tag{2.2}$$

Here, A, is the surface area of the undisturbed surface. Technically, since we have assumed a semi-infinite surface, A is infinite. But, we may imagine taking a long, finite, surface to make A finite, and then taking a limit. We ignore this complication. We further simplify the surface energy by approximating the radical via

$$\sqrt{1 + u'^2} \approx 1 + \frac{u'^2}{2}. \tag{2.3}$$

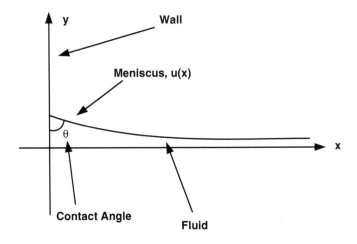

FIGURE 2.5: The geometry for our simplified meniscus calculation.

This is, of course, only valid for small deflections of the surface. With this approximation, the surface energy reduces to

$$\sigma \int_0^\infty \frac{u'^2}{2} dx. \tag{2.4}$$

The gravitational energy of our system may be written as

$$\rho g \int_0^\infty \int_0^{u(x)} y \, dy \, dx. \tag{2.5}$$

Here, ρ is the fluid density, g the gravitational constant, and the zero of gravitational potential has been defined to be the x-axis. The integration with respect to y may be carried out and hence the gravitational energy written as

$$\rho g \int_0^\infty \frac{u^2}{2} dx. \tag{2.6}$$

Combining the surface and gravitational energies, we may write our total energy as

$$E[u(x)] = \int_0^\infty \left(\sigma \frac{u'^2}{2} + \rho g \frac{u^2}{2} \right) dx. \tag{2.7}$$

Equation (2.7) is an example of a *functional*. The input is a function, $u(x)$, and the output a real number. $E[u(x)]$ defines a map between some set of admissible functions and the real numbers. Our claim is that nature chooses the meniscus that minimizes this energy, i.e., nature chooses the function

$u(x)$ to make $E[u(x)]$ as small as possible. The theory of the minimization of functionals is the subject of the *calculus of variations*. The reader is referred to Appendix A for a brief introduction to this subject, or to the Related Reading section at the end of this chapter for a more complete presentation.

Here, we simply make use of the necessary Euler-Lagrange equation. For integrals of the type

$$\int_a^b F(x, u, u') dx \tag{2.8}$$

the Euler-Lagrange equation is

$$\frac{\partial F}{\partial u} - \frac{\partial}{\partial x} \frac{\partial F}{\partial u'} = 0. \tag{2.9}$$

That is, in order for $u(x)$ to be a minimizer of Equation (2.8) it is necessary that $u(x)$ satisfy Equation (2.9). Applying this to Equation (2.7) we obtain the second order ordinary differential equation

$$\frac{d^2 u}{dx^2} - \frac{\rho g}{\sigma} u = 0. \tag{2.10}$$

This equation is easily solved. We find

$$u(x) = c_0 \exp(-\sqrt{\frac{\rho g}{\sigma}} x) + c_1 \exp(\sqrt{\frac{\rho g}{\sigma}} x). \tag{2.11}$$

To completely determine the meniscus shape, we must impose boundary conditions on this solution. We require that the disturbance to the interface goes to zero as x goes to infinity. That is, at infinity, we require that the interface return to its undisturbed state. This requires that we choose $c_1 = 0$. The second unknown constant, c_0, must be determined from a condition at the wall. In the early 1800's, Thomas Young introduced the notion of a *contact angle* condition. The contact angle condition says that the surface of a liquid makes contact with a solid at a fixed angle. This angle depends on the properties of the liquid and the solid. Measuring the contact angle as shown in Figure 2.5, we may write the contact angle condition for our problem as

$$u'(0) = -\cot(\theta). \tag{2.12}$$

Hence, the meniscus shape is given by

$$u(x) = \sqrt{\frac{\sigma}{\rho g}} \cot(\theta) \exp(-\sqrt{\frac{\rho g}{\sigma}} x). \tag{2.13}$$

Profile - Leonard M. Adleman

The movie *Sneakers* starring Robert Redford, Sidney Poitier, Dan Aykroyd, and River Phoenix, contains a scene where a scientist lectures on a breakthrough in cryptography. A small credit at the end of the film lists Leonard M. Adleman as a "mathematical consultant." In fact, as recounted on his web page, Adleman actually wrote the dialog and designed the slides of equations used in that scene. Adleman notes that he did not win an Academy Award for this work. It is unclear as to who received the award for best mathematical consultant that year.

While it does seem unlikely that Adleman will be long remembered for his movie work, it is certain that he will be remembered as the father of *DNA Computing*. Adleman single-handedly launched the field of DNA Computing with his 1994 article in *Science* [1], titled *Molecular Computation of Solutions to Combinatorial Problems*. Interestingly enough, he begins with a reference to Feynman:

> In 1959, Richard Feynman gave a visionary talk describing the possibility of building computers that were "sub-microscopic." Despite remarkable progress in computer miniaturization, this goal has yet to be achieved. Here, the possibility of computing directly with molecules is explored.

Adleman goes on to describe a new approach to solving computationally difficult problems, such as the classic directed Hamiltonian path problem. Using DNA and PCR amplification, Adleman actually carried out the solution of an instance of this problem, demonstrating the feasibility of computing with DNA. While self-assembly is a young field, this work will undoubtedly be viewed as a landmark, not only affecting subsequent research in DNA computing, but in the use of DNA in engineered systems in general.

Now, if Adleman is somehow *not* remembered for either his movie work or the invention of DNA computing, he still has a fall back position. Leonard M. Adleman is the "A" in the well-known RSA public-key cryptography system and one of the three co-authors of the seminal paper proposing this scheme. The RSA scheme is the primary cryptographic scheme used on the internet and world wide web.

Leonard M. Adleman was trained as a computer scientist, receiving both a bachelor's degree and doctorate from the University of California at Berkeley. Presently, he is the Henry Salvatori Professor of Computer Science at the University of Southern California. He continues to be heavily involved in cryptography, DNA computing, and more recently, theoretical aspects of self-assembly.

This solution contains a parameter, L, known as the capillary length

$$L = \sqrt{\frac{\sigma}{\rho g}}. \qquad (2.14)$$

In terms of L we may write the meniscus shape as

$$u(x) = L \cot(\theta) \exp(-x/L). \qquad (2.15)$$

Note that L provides a measure of the distance over which the distortion due to the wall is felt by the liquid. When we have moved a distance L away from the wall, the maximum meniscus distortion has decayed by a factor of e. Also note that the contact angle, θ, determines the shape of the meniscus. If the contact angle satisfies $0 < \theta < \pi/2$, we say that the liquid *wets* the surface and the meniscus appears as in Figure 2.5. If the contact angle satisfies $\pi/2 < \theta < \pi$, we call the surface *non-wetting*, and the meniscus appears as in Figure 2.6.

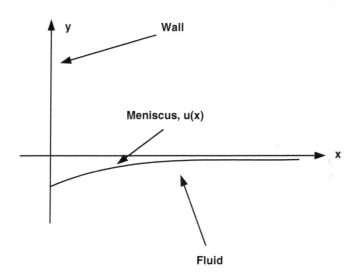

FIGURE 2.6: The shape of the meniscus for a non-wetting system.

2.2.3 Back to the Bubble Raft

Now that we understand why the surface of a liquid becomes distorted in the presence of a solid, we can easily understand why the bubbles in our bubble raft clump together and are attracted to the walls of the container.

The later situation is easiest to understand. Consider the bubble-meniscus system shown in Figure 2.7. The bubble has risen to the surface of the fluid

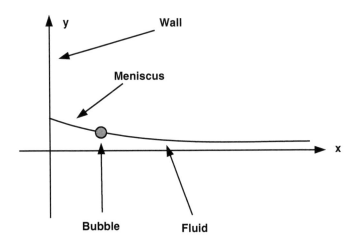

FIGURE 2.7: A bubble floating on a fluid near a wall. The bubble feels a buoyant force in the y-direction and hence moves up the meniscus towards the wall. Hence, it appears that the bubble is attracted to the wall.

because it is buoyant. The density of the bubble is less than that of the liquid, hence it feels a net upward force. Once it reaches the surface of the fluid, it still feels a net upward force, but is now constrained to live on the surface. The only way for the bubble to continue its upward journey is to move along the meniscus, towards the wall. Hence, the bubble approaches the wall and it appears to us as if there is a force of attraction between the bubble and the wall. Notice that we have assumed that the liquid wets the wall. If the system is non-wetting and the meniscus appears as in Figure 2.6, the bubble will move up the meniscus and away from the wall. It will appear as if there is a repulsive force between the bubble and the wall.

In the bubble-wall example, we ignored the fact that the bubble also distorts the surface of the liquid and produces its own meniscus. While this is negligible in the bubble-wall example, if we want to understand bubble-bubble interactions, we must consider the bubble's meniscus. A pair of bubbles floating at the surface of a liquid will appear as in Figure 2.8. Even though the meniscus between the bubbles is curved upwards, if the bubbles move towards one another, they will be moving down the menisci overall and reducing the gravitational energy in the system. Hence, it appears as if there is an attrac-

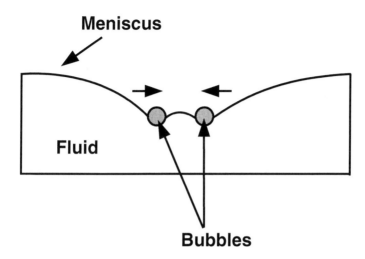

FIGURE 2.8:   Two bubbles floating on the surface of a fluid. Motion towards one another reduces the gravitational energy of the system, hence there is an apparent attractive force between the bubbles.

tive force between the bubbles.

In Chapter 5, we will revisit the meniscus effect and consider capillary forces between objects other than bubbles. One can imagine that a variety of situations are possible, leading to both attraction and repulsion. The reader is also referred to the excellent review papers by Kralchevsky and Nagayama [77] and by Mansfield et al. [89]. The more recent paper by Vella and Mahadevan [133] is also recommended. But, we still have not completed our explanation of the bubble raft. We now understand why bubbles are attracted to one another, but a glance at Figure 2.2 indicates that more is happening than just attraction. The bubbles in the raft do not stick together randomly, rather there is a short and long range order to the bubble raft.

To understand the origins of this order, we must understand how objects *pack.* You are probably familiar with the classic grocer's problem of how best to stack a collection of identical oranges. This problem has a rich history in mathematics and physics, starting with Kepler in the 1600's and culminating with the work of Hale in 1998. An excellent account of this story and other packing problems may be found in the book by Aste and Weaire [8].

Here however, we are interested in how objects pack in two dimensions rather than three. As we've already seen, our bubbles are constrained to move on the surface of the fluid and move towards one another in order to reduce the overall energy of the system. So, if we have three bubbles, as in Figure 2.9 (a), these three bubbles will tend to move towards each other. One possibility is that they take up the configuration in Figure 2.9 (b). However,

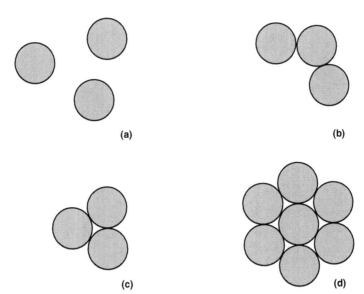

FIGURE 2.9:   Different configurations of bubbles on the surface of a fluid.
(a) Three separate bubbles. Capillary forces will draw these bubbles together.
(b) One possible packing of three bubbles. (c) A lower energy packing of three
bubbles. (d) A hexagonal optimal bubble packing.

in this configuration it is clear that the energy can be reduced further. The
bubbles have formed a chain. The bubbles on the end have minimized gravitational potential with respect to the middle bubble, but not with respect
to each other. If they roll around the edge of the middle bubble, they can
reduce the gravitational potential between each other without increasing the
potential interaction with the middle bubble. Eventually, they'll reach the
configuration of Figure 2.9 (c). Additional bubbles will join the cluster in the
same way, eventually resulting in the hexagonal structure of Figure 2.9 (d).
When this structure is repeated throughout the plane we have the hexagonal
close packing structure of Figure 2.2. It is well known that this packing is
the optimal packing for disks in a plane. In this configuration, the so-called
packing fraction, i.e., the area of the plane covered by bubbles divided by
the total area, is approximately 0.91.[4] Finally, we note that the presence of
the hexagonal close packing arrangement relies on the fact that the bubble
tray is effectively infinite. The edges do not constrain or effect the packing
in any way. Edge effects can dramatically change the structures that appear.
This is explored in the exercises for this chapter and will be discussed again
in Chapter 6.

2.3   Crystallization

When I was young, one of my favorite experiments was to grow salt crystals. My brothers and I would warm a pot of water, dissolve as much salt as we could in the pot, and pour the result in to all manner of scavenged jars, cups, and cans. We'd then place these on top of radiators near windowsills all over the house and wait for the water to evaporate, leaving behind a crust of salt crystals. Eventually, my mother would collect the dishes and either throw them out or wash them, setting back the advance of our scientific endeavors immeasurably. Mostly, the crystals we grew would be no larger than the original salt crystals we began with, but sometimes, we'd get lucky and manage to grow a large cubic crystal about the size of a fingernail.

If we could have examined the atomic structure of a salt crystal we might have conjectured that the regular shape we saw at the macro-level was due to the organized, periodic arrangement of atoms at the atomic level. In fact, this is the very definition of a crystal. A *crystal* is a solid in which the atoms or molecules are arranged in a regular, periodic array. Since this array is repeated throughout the crystal, we can think about crystals in terms of their *unit cells*. This is the smallest structural unit of the crystal and just as with the hexagonal structure of the bubble raft, is repeated throughout the entire crystal.

For salt, the unit cell is a cube made up of both sodium and chloride ions arranged in a three-dimensional lattice such as shown in Figure 2.10. Other ionic crystals exhibit a different atomic order. This order can be quite complex for ionic crystals simply because multiple ionic species must fit together in

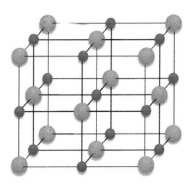

FIGURE 2.10:   The unit cell of NaCl, salt. The large spheres represent chloride ions, the small spheres represent sodium ions.

some regular manner. Metals, on the other hand, are comparatively simple.

A pure metal is a crystalline solid made up of atoms of all the same species. Hence, the atomic structure of metals may be understood in terms of sphere packing. As we saw with the bubble raft, the best way to pack spheres in a plane is the hexagonal close packing arrangement of Figure 2.9. We can now imagine growing a metallic crystal by adding layers on top of these close packed spheres. We are back to the grocer's packing problem! But, there are two ways to proceed. The grocer would add the second layer by adding a new sphere on top of each of the voids in the first layer. This leads to the typical orange or cannonball sphere packing now known to be the tightest possible packing in three dimensions. Nature, however, sometimes proceeds differently. Instead of laying spheres on top of the voids sometimes spheres are placed on top of spheres from the lower level. The hexagonal planar close packing structure is simply repeated. Alternatively, nature sometimes begins with a different two-dimensional packing, such as a simple square lattice, obtained by laying each sphere on top of the vertices of a grid. This pattern is then repeated as the crystal extends into three dimensions.

FIGURE 2.11: A scanning electron microscope photograph of a four-micron size iron crystal. This is from the Apollo 15 Hadley-Apennino lunar landing site. Credit: NASA Johnson Space Center.

From the point of view of self-assembly, what is of interest is not only the various structures that may form, but how they form, i.e., the process of crystallization. A simple analogy to help us understand crystallization is possible. Imagine we lay out a sheet of bubble wrap on the floor. We'll imagine that each of the bubbles in the bubble wrap is an atom or molecule and that this represents the outer surface of some crystalline structure.

Try It Yourself - The Bubble Raft

A bubble raft is fairly easy to create and is well-worth the minimal time and effort needed. To carry out this experiment yourself you will need:

Materials

- Beaker, Petri-dish, or other large-mouthed container.

- Soap solution. A mix of 12 parts water to one part standard dish soap works well. Make sure the dish soap is free of additives such as anti-bacterial agents or scents.

- A small reciprocating pump. These may be purchased at almost any pet store. Look for a cheap pump intended for an aquarium.

- Power source for the pump. Batteries work well, but a variable power supply will give you more control over bubble size.

Procedure Pour a small amount of soap solution into your Petri-dish. Secure the output of the pump beneath the surface of the liquid. Attach the pump to your power supply and adjust the voltage until you obtain bubbles of the size you desire. Bubbles approximately 1mm in diameter work well. If the outlet of your pump is too large try attaching a nozzle with a small hole to the outlet tube. An inexpensive solution is to place a piece of tape over the end of the outlet tube and poke a small hole in the tape. Now, simply allow the system to run until the bubble raft is formed. If you are having difficulty with bubbles clumping near the pump output, try placing a small fan so that it blows across the surface of the liquid, gently drawing the bubbles away from the pump.

Things to Try

- Once your raft has formed, take a straw and gently agitate the bubble raft solution. Once the bubble raft is broken apart, allow it to re-assemble. How does this happen? How long does this take?

- Identify defects in your bubble raft. Tilt or gently tap your Petri-dish. What happens to the bubbles near a defect?

- Create a bubble raft using two pumps. Adjust the output so that two different size bubbles are created. What sort of structure is formed? How does this vary as the ratio of bubble radii varies?

Further Reading The text, *Teaching General Chemistry: A Materials Science Companion*, by A. Ellis et al. contains a description of a more elaborate bubble raft experiment and numerous ideas for experimentation with bubble rafts. See also Appendix B for related web pages.

To add the next layer, we drop ping-pong balls onto the bubble wrap.[5] When a ball hits the bubble wrap it bounces around, eventually finding a resting place nestled in the void created by the bubbles. As additional balls are added they may jostle and knock other balls into different configurations. Occasionally, a ping-pong ball may balance on the top of a bubble, but as the process continues it is certain to be knocked out of this local energy equilibria and into the lower energy state of a void. Proceeding in this way we eventually cover the sheet of bubble wrap; we've formed a crystalline layer of ping-pong balls, and can continue the process.

Of course, the bubble-wrap analogy is highly idealized. It is possible that not all voids are filled in a given layer or that the occasional tennis ball makes its way into the structure. It might be that the tops of the bubbles are more energetically favorable than the voids and that we manage to self-assemble a different sort of crystalline structure. Another, even more likely possibility, is that we don't get to start with a single sheet of bubble wrap oriented perfectly on the floor. If we did, we would eventually form a pure crystal with a nice large scale structure like the crystal of iron in Figure 2.11. But, this only happens if the crystals are allowed to grow very slowly or under very controlled conditions. At the macro-scale, what we usually think of when

FIGURE 2.12: A large crystalline structure. Note the multiple facets.
Credit: iStockPhoto.com/Steve Goodwin.

we think of crystals are structures such as those shown in Figure 2.12 or in Color Plate 11.4. The structure of the bismuth crystal of Color Plate 11.4 is especially remarkable and intricate. How do these structures arise?

If we return to the bubble wrap analogy, what we start with is not a single piece of nicely oriented bubble wrap, but rather with several pieces of bubble wrap that have assembled individually and are randomly arrayed with respect to one another. This is the start of what is called a *polycrystal*. A layer of a polycrystal would appear as in Figure 2.13. As we grow our polycrystal

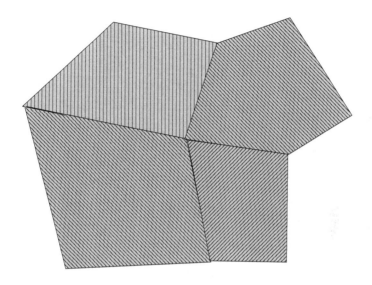

FIGURE 2.13: A cross-section of a polycrystal. The lines indicate the orientation of the lattice.

each of the individual bubble wrap layers nucleates nicely, but the layers grow together in an irregular manner each one leading to a facet of the macro-scale polycrystal. If we force the process to occur even more rapidly, and allow less time for the individual atoms to rearrange themselves into a perfect structure, we end up with a crystal like the ice crystal of Figure 2.14 or the multi-colored fluoride crystal of Color Plate 11.6.

Now, when we studied bubble rafts, it was useful to construct a mathematical model in order to gain insight into how the raft self-assembled. For large self-assembling systems, like crystals, constructing a simple analytically treatable model is not possible. In order to understand large systems of interacting particles it is useful to construct computational models. As our first example of such a model, we consider *Diffusion Limited Aggregation* (DLA) which was introduced by Witten and Sander [143, 144] in 1981 as a general model of aggregation processes and dendritic growth. The idea behind DLA is simple. We begin with some computational domain, typically a large square lattice. At the center of the lattice a fixed seed particle is placed. A second particle is then added to the system at a random location far away from the origin. This particle undergoes a random walk on the lattice until it encounters a site

FIGURE 2.14: Ice crystals growing on a window. Note the dendritic structure. Credit: iStockPhoto.com/Jamie Farrant

adjacent to the seed. Once it does, this walker is fixed in place and becomes part of the growing cluster. Additional walkers are then added, the process repeated and the cluster allowed to grow.

A DLA simulation typically leads to a structure such as the one shown in Figure 2.15. This is clearly dendritic and does begin to resemble crystals such as those in Figure 2.14 or Color Plate 11.6. Even more remarkably, DLA simulations have been used to model electrodeposition, viscous fingering in the Hele-Shaw instability, and even the growth of bacterial colonies. Numerous researchers have modified the DLA model, introducing variations such as non-square grids, particles sticking with a probability, curvature dependent local particle behavior, and global constraints. The DLA model has also been shown to produce objects with a self-similar, fractal, geometry. An excellent recent review article was written by Sander [115], one of the two originators of the DLA model. The richness of the DLA model is somewhat surprising. Even though much progress has been made, after more than twenty years of research the model has resisted an exact analytic treatment. From the point of view of self-assembly this is both encouraging and discouraging. It is certainly encouraging to discover that even very simple assembly algorithms can lead to rich and complex behavior. However, the difficulties inherent in understanding even such a simple model should give us pause, lest we claim too much about what we understand about how nature utilizes self-assembly.

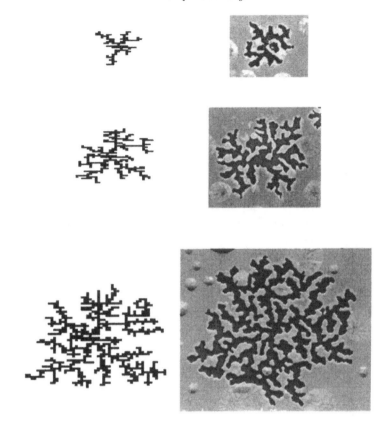

FIGURE 2.15: A typical DLA simulation (left) and experimental system (right) exhibiting DLA type growth. Reprinted with permission from Kwon, et al., J. App. Phys., v. 93, pp. 3270-3278, Copyright 2003, American Institute of Physics.

2.4   Polymerization

If it is almost impossible to go through a day without encountering a bubble, it certainly is impossible to go through a day without encountering a polymer. The plastic case of your cell phone, the synthetic fibers in your pants, the rubber eraser on the tip of your pencil, and the many many proteins present in your body are all examples of polymers. Polymers are large molecules known as *macromolecules* that in their simplest form result when many identical short subunits, *monomers*, bind together in a long repeating chain. A simple example of such a polymer is polyethylene. In polyethylene, the monomer, or subunit, is C_2H_4, commonly known as ethylene. The chemical structure of ethylene is shown in Figure 2.16. When ethylene is poly-

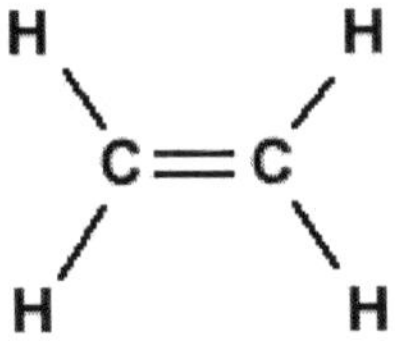

FIGURE 2.16: The chemical structure of ethylene. The two CH_2 subunits are held together by a double bond between the carbon atoms.

merized, a carbon-carbon bond in the subunit is broken and a carbon-carbon bond is formed between adjacent subunits. This results in the polymer shown in Figure 2.17. The number of repeating units in a polymer is known as the *degree of polymerization*.

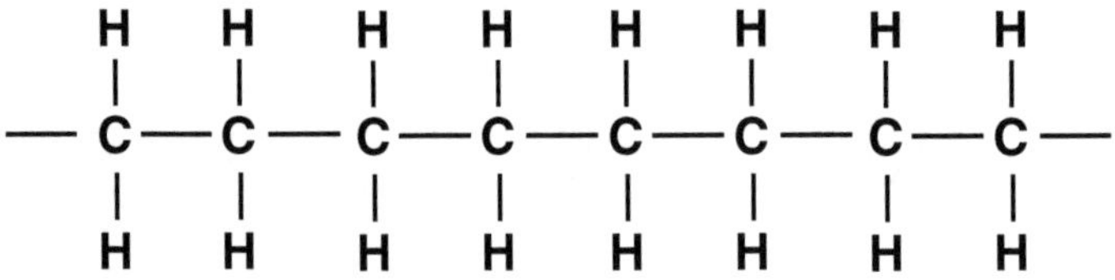

FIGURE 2.17: The chemical structure of polyethylene. Only a short segment of repeating CH_2 units is shown. In a real polymer, the chain would typically be several thousand units long.

In addition to simple linear chains comprised of monomers, a variety of more complex polymer structures are possible. *Branched polymers* form when short chains of monomers are attached to the main polymer chain. This structure is shown in Figure 2.18. *Cross-linked polymers* form when the side branches of branched polymers become attached to other polymer chains. This structure is shown in Figure 2.19. Polymer structure becomes even more complex when more than one type of monomer is present in the polymer chains. These are referred to as *copolymers*. Various structures are possible. In *random copolymers*, a linear chain is formed from multiple monomers with no regard to order along the chain. In *block copolymers*, short chains of the same type of monomer alternate along the chain. In *alternating copolymers* monomer units alternate, one by one. For a more detailed discussion of the structure and properties of polymers the reader is referred to the Related Reading section at the end of this chapter.

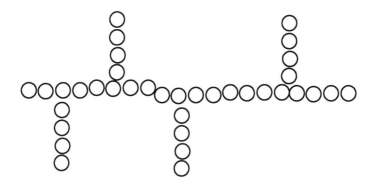

FIGURE 2.18: The basic structure of a branched polymer. Each circle represents a monomer.

Here, what is of interest to us is the process of polymerization. Polymerization is a simple example of a self-assembly process. Individual particles, monomers, placed in the right environment, spontaneously arrange themselves into organized, complex, structures. Next, let's examine how this process might be modelled, and what the analysis of such a model might tell us. This model is adapted from [11].

Let's consider the simplest possible system, a well-stirred bath containing a single species of monomers or subunits. When n of these subunits are strung together in a chain, we'll call this a polymer and denote the polymer of length n by P_n. The single monomers are then denoted P_0. Polymerization typically requires the presence of a second species that does not become bound to the polymer, but allows the process of polymerization to occur. For example, in the case of the ethylene subunits of Figure 2.16, a second molecule is needed to break the double carbon bond and allow the subunits to join and form the structure of Figure 2.17. Here, we'll denote this second species by A, and assume the polymerization reaction proceeds according to

$$A + P_n \longrightarrow P_{n+1}. \tag{2.16}$$

Note that A is used up in this reaction. In order to mathematically describe this process, we use the *Law of Mass Action*. This law states that the rate of our chemical reaction is proportional to the concentrations of the molecules participating in the reaction. We denote the concentration of A by $a(t)$ and of P_n by $p_n(t)$. Then, the Law of Mass Action implies

$$\frac{dp_0}{dt} = -kap_0 \tag{2.17}$$

$$\frac{dp_n}{dt} = ka(p_{n-1} - p_n) \tag{2.18}$$

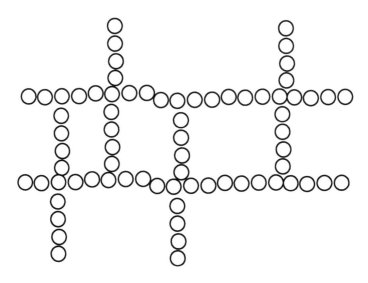

FIGURE 2.19: The basic structure of a cross-linked polymer. Each circle represents a monomer.

$$\frac{da}{dt} = -ka \sum_{n=0}^{\infty} p_n. \qquad (2.19)$$

The constant of proportionality, k, is called the *rate constant* for the reaction. We'll impose the initial conditions that $a(0)$ and $p_0(0)$ are nonzero and that $p_n(0)$ is zero for $n \geq 1$. Now, we'll also assume that no monomers are added to the tank during the process and that no monomers or polymers are removed. This implies that we have conservation of overall monomer concentration. This may be written as

$$\sum_{n=0}^{\infty} p_n(t) = p_0(0). \qquad (2.20)$$

Equation (2.19) is easily solved for $a(t)$. We find

$$a(t) = a(0)\exp(-kp_0(0)t). \qquad (2.21)$$

We may now eliminate $a(t)$ from the equations for $p_n(t)$. It is convenient to define a new variable $y(t)$ such that $y' = ka(t)$ and $y(0) = 0$. We may then rewrite Equations (2.17)-(2.18) as

$$\frac{dp_0}{dy} = -p_0 \qquad (2.22)$$

$$\frac{dp_n}{dy} = p_{n-1} - p_n. \qquad (2.23)$$

This system is now easily solved. We find

$$\frac{p_n(y)}{p_0(0)} = \frac{y^n e^{-y}}{n!}.$$

(2.24)

This gives us a distribution of polymer chain lengths as a function of y. This particular distribution is known as a *Poisson distribution*. The mean and variance are both equal to y. Now, y, is a function of time, hence this distribution gives us the polymer distribution at an instant in time. The fact

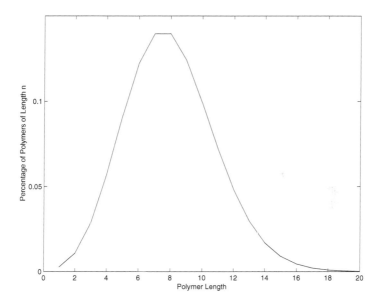

FIGURE 2.20:   The distribution of polymers in an evolving polymerization reaction for the model of Section 2.4.

that we have a distribution of polymer lengths indicates that this form of self-assembly is fundamentally different than that for the bubble raft. At the end of the process, we do not produce one fixed uniform product. Rather, we always produce a range of end products. As we'll see in Part II, this is typical of self-assembling systems and presents the engineer with a difficult design problem.

2.5 Micelles

As our final example of a naturally occurring inorganic self-assembling system, we consider the *micelle*. The micelle is a molecular structure, formed in solution, from individual *lipid* molecules. Lipids are *amphiphilic* molecules. They are both *hydrophillic*, or water-loving, and *hydrophobic*, or water-hating. The head group of a lipid is the hydrophillic end and is typically a polar molecule soluble in water. The tail of a lipid is hydrophobic and is typically composed of nonpolar hydrocarbons. A typical lipid structure is shown in Figure 2.21.

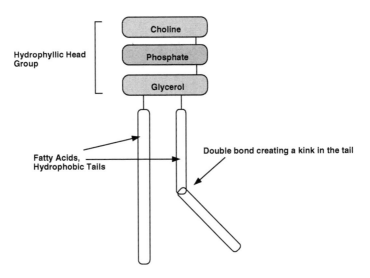

FIGURE 2.21: A typical lipid molecule. This schematic shows phosphatidylcholine, but the general structure holds for most lipids. The kink in the tail influences the molecule's ability to pack with its neighbors.

The schizophrenic nature of lipids leads to remarkable behavior when the lipid molecules are placed in water. Imagine adding a single lipid molecule to a glass of water. The hydrophillic head, being readily soluble, will be quite "happy" in solution. On the other hand, the hydrophobic tail, will be quite unhappy and will attempt to move the lipid away from water molecules. If there is only one lipid in the solution, or a very low concentration of lipids, the easiest way to energetically satisfy both the water-loving and water-hating desires will be for the lipid to migrate toward the surface and situate itself like an ostrich with its head in the sand; the head group stays in the water, the tail pokes through the surface. When enough lipid molecules are placed in the solution, eventually, the *critical micelle concentration* is achieved. At this point it becomes energetically favorable for the lipids to interact with one another in order to satisfy their amphiphilic desires. The lipids begin to form *micelles*. The simplest micellar structure is shown in Figure 2.22. We see that

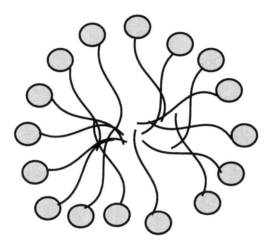

FIGURE 2.22:   A cross-section of a lipid micelle.

in some aqueous solution, the micelles have clumped together such that their tails are adjacent, and hence avoid the water, while their heads are still in the aqueous solution.

But, wait a minute. Why should the lipids form a spherical micelle? A moments thought is enough to convince oneself that many other structures are possible. The lipids could arrange themselves in a long planar bilayer with the head groups facing outwards and the tails facing inwards. This bilayer could perhaps bend back on itself and form a sphere with water on the inside as well as the outside. Or, the lipids might group together and form a cylinder. Or, the heads might group together around small globules of water, while the tails, having forced the water out happily exist in some loose super-matrix. In fact, all of these possibilities are realizable.

Which configuration the lipids choose depends on the relative size of the head groups and tails. Just as with the self-assembly of bubble rafts and crystals, the notion of packing comes into play. Israelachvili [66] introduced the notion of the semi-empirical critical packing parameter in order to quantify the formation of different micellar structures. This packing parameter is typically written as

$$\frac{v}{a_0 l_c} \qquad (2.25)$$

where v measures the volume of the hydrocarbon chain, l_c measures the maximum effective length that the chains can assume, and a_0 measures the surface area occupied by the head group in a given micellar structure. The structures formed in a given solution are characterized in terms of this ratio. The possibilities are shown in Figure 2.23. We see that when the critical packing

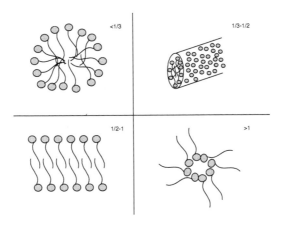

FIGURE 2.23: Possible structures formed by lipid micelles. The value of the packing parameter, indicated in the upper right-hand corner of each diagram, characterizes the phase selected.

parameter is small, meaning our lipids have large heads and slender tails, the lipids form spherical micelles. When the heads and tails become roughly the same size, the lipids tend to form bi-layers and when the tails become larger than the heads the system enters the reverse micelle phase. The reader is referred to the text by Israelachvili [66] and by Larson [79] for a more detailed discussion of the micelle phase diagram. The review article by Nagarajan and Ruckenstein [95] gives a detailed introduction to the thermodynamics of micelle formation. The note by Smit et al. [124] contains a nice discussion of computational models of micelle formation.

Finally, we note that while micelles are interesting from the point of view of self-assembly, they also hold tremendous technological promise. At the macrolevel, the presence of micelles in a solution effects the fluid rheology in a variety of ways [79]. In the microworld, the biomimetic properties of micelles makes them ideal for both attempting to understand how biological membranes function and for duplicating the behavior of biological membranes.

Two achievements in this direction warrant special mention. In 1992, Bachmann et al. [9] created a micelle capable of transporting an organic molecule across its boundary. Once inside the micelle, the organic molecule undergoes hydrolysis, producing more membrane material. This membrane material migrates to the boundary and enlarges the micelle. Eventually, the micelle becomes unstable and divides into two smaller micelles. This effectively mimics cellular growth and division. In a different sort of study, Rosler et al. [105] showed how micelles could be used to encapsulate drugs as part of a targeted drug delivery system. They further showed that the micelle membrane could be made functional, in much the same way that a cell membrane is functional, by embedding proteins in the membrane to create channels.

2.6   Chapter Highlights

- Four examples of inorganic self-assembling systems were presented in this chapter. They include the *bubble raft*, *crystals*, *polymers*, and *micelles*.

- The *bubble raft* self-assembles under the action of capillary forces. The fully assembled system exhibits short and long range order. This order may be understood in terms of optimal packing theory.

- *Crystals* are naturally occurring self-assembled systems. They exhibit atomic scale order and macroscale structure. The atomic scale order may also be understood in terms of packing theory. The macroscale structure is highly dependent on the process of crystallization.

- *Polymers* are long-chain macromolecules comprised of small individual subunits called *monomers*. The polymerization reaction is an example of a self-assembly process. This process leads to a distribution of polymer chains of varying lengths rather than a uniform output.

- The *micelle* is formed from individual amphiphilic molecules placed in solution. The micelle structure may be controlled by varying the size of the hydrophillic head group. Micelles are closely related to biological membranes.

2.7 Exercises

Section 2.2

1. Find the 1947 paper by Bragg and Nye [21]. This paper contains sixteen pictures of bubble rafts. Why? Explain why each picture was included in this paper and what feature of crystal behavior each illustrates.

2. In 1956 the British fluid dynamicist, G.I. Taylor, was awarded the de Morgan medal of the London Mathematical Society. In his address to the society, he noted *If a solid sphere - say a wooden ball - floats on the surface of a fluid which is rotating like a solid body about a vertical axis, it will gravitate towards the lowest point, which is on the axis of rotation. If however, the sphere be weighted by fixing a piece of lead to its surface till it nearly sinks the sphere will travel outward and therefore upwards till it strikes the wall of the vessel containing the fluid.* Explain why this happens. How is this like the meniscus effect? How is it different?

3. Compute the shape of the meniscus for a plate placed in a semi-infinite fluid at an angle. How does the meniscus vary as the angle of the plate is varied?

4. Wax does not wet water. The contact angle between wax and water for the plate-water system is approximately 107°. Compute the shape of the meniscus for this system and plot your result.

5. The hexagonal packing of the bubble raft occurs because the system is effectively unconstrained by the container. When boundaries are present, other arrangements may be optimal. To illustrate this, obtain a few hundred pennies and attempt to pack them into various shapes drawn on a piece of paper. Try various size circles, rectangles, and squares. What do you observe? Can you prove anything about the optimality of your packing for various shapes?

Section 2.3

6. Diffusion limited aggregation (DLA) is easily simulated on the computer. A simple 2-d model of DLA has been implemented in MIT Media Lab's Star Logo. Download the implementation of Star Logo appropriate for your computer and explore the DLA simulation. What aspects of DLA have been built into this simulation? What has been left out? What patterns can be formed by changing the parameters in this simulation? How would you modify or improve this model?

7. This problem requires a large group of people. Give each person in the group a pad of sticky notes and attach one note to the center of

a wall or blackboard. Now, have each person in the group add a note to the board following the rule that new notes must be placed adjacent to existing notes. Continue this process. What structure forms? Can you modify the rule in a simple way such that the structure your group grows becomes more regular?

8. The DLA model leads to dendritic structures. Why? Why do fissures appear and persist? Why are they not filled in by the random walker process?

9. Crystals often contain *dislocations*. One particularly interesting dislocation is the *screw dislocation*. Explain the origin of the screw dislocation and the macroscopic crystalline forms that arise when a screw dislocation is present.

Section 2.4

10. Return to our simple model of polymerization and explicitly solve for y as a function of time. At what time should one stop the reaction to obtain the maximum number of chains of a given length, l?

11. Modify the model of this section by allowing the reaction rate, k, to depend on polymer length, n. What happens to the distribution when k increases with n? What happens when k decreases with n?

Section 2.5

12. Soap films are composed of lipid molecules. What is the structure of a soap film? How are the lipid molecules arranged? Where is the water? For a soap film surface, the surface tension is actually twice what we consider to be the surface tension of the bulk fluid. Why?

2.8 Related Reading

A rigorous mathematical introduction to capillarity can be found in the nice book by Robert Finn:

R. Finn, *Equilibrium Capillary Surfaces*, Springer-Verlag, 1986.

A somewhat simpler introduction to the mathematics may be found in:

J. Oprea, *The Mathematics of Soap Films: Explorations with Maple*, American Mathematical Society, 2000.

The classic text of Batchelor contains an excellent discussion of surface tension and the Laplace-Young law:

G.K. Batchelor, *An Introduction to Fluid Dynamics*, Cambridge University Press, 1967.

Isenberg's text also contains a nice discussion of surface tension and the Laplace-Young law and makes for entertaining reading as well. He also gives a short, informal, introduction to the calculus of variations.

C. Isenberg, *The Science of Soap Films and Soap Bubbles*, Dover, 1992.

The charming book of Boys is a must have for anyone interested in capillarity or soap bubbles:

C.V. Boys, *Soap Bubbles: Their Colors and Forces Which Mold Them*, Dover, 1959.

If you are interested in a rigorous introduction to the calculus of variations, the text by Gelfand and Fomin is the place to start:

I.M. Gelfand and S.V. Fomin, *Calculus of Variation*, Dover, 2000.

The text on packing by Aste and Weaire is thoroughly enjoyable.

T. Aste and D. Weaire, *The Pursuit of Perfect Packing*, Institute of Physics Publishing, 2000.

Although the focus is not on self-assembly per se, the book by Philip Ball discusses much that is related to what has been discussed in this chapter. It is easily readable and the sections on crystallization, surface tension, and colloids are especially interesting.

P. Ball, *Designing the Molecular World*, Princeton University Press, 1994.

The book by Israelachvili, although only twenty years old, has become the classic text on surface forces. The third part of the text focuses on self-assembly of micelles and related structures.

J. Israelachvili, *Intermolecular and Surface Forces*, Academic Press, 1992.

The formation of micelles in a fluid endows that fluid with all sorts of unusual properties. This class of fluids is called *complex fluids*. The wonderfully comprehensive text by Larson explores just about every possible type of complex fluid.

R.G. Larson, *The Structure and Rheology of Complex Fluids*, Oxford University Press, 1999.

2.9   Notes

1. At the beginning of their 1947 article, Bragg and Nye criticize the then standard macroscopic models of crystal structure. They discuss magnets floating on the surface of a liquid and circular disks floating on a liquid and held together by capillary attraction. As we'll see in Part II, both of these ideas have been resurrected by researchers in self-assembly, often to great effect.

2. See the Related Reading section for a complete derivation of the Laplace-Young Law. The texts by Oprea, Isenberg, and Batchelor are especially useful.

3. The equation $H = 0$ is called the *minimal surface equation*. Some of the most interesting and difficult mathematics of the last century has centered around this equation. The equation $H = 1$ is called the *constant mean curvature equation*. It too has been the source of much fascinating mathematical work. See the Related Reading section for further information on these topics.

4. In fact it is exactly equal to $\pi/\sqrt{12}$. See [8] for a proof.

5. Of course, we're imagining we are using the bubble wrap that has *large* bubbles. If you imagine the kind with small bubbles, you'll have to adjust your ping-pong balls accordingly.

Chapter 3

Organic Systems

The human understanding, from its peculiar nature, easily supposes a greater degree of order and equality in things than it really finds.

Bacon, *Novum Organum*, I, 45

3.1 Introduction

The self-assembly of bubble rafts, crystals, polymers, and micelles is quite impressive. But, these inorganic systems pale in comparison with nature's *organic* self-assembled structures. In this chapter we examine three biological systems that are known to self-assemble and that are at least partially understood. The tantalizing possibility of using self-assembly to create and control intricate functional structures such as the biological structures discussed here is the main reason for the level of excitement surrounding self-assembly today.

We begin in Section 3.2 with a discussion of *proteins* and the *protein folding problem*. Proteins are polymers; long chain molecules composed of simple subunits known as amino acids. These basic structures are the most important building block for the cell, and consequently for all of biology. In the last chapter, we studied polymerization; the process by which long chain molecules are constructed. Here, we focus on what happens after the chain has been built - the protein folding problem. It is the ability of proteins to uniquely fold into a wide variety of shapes that makes them such useful nanoscale machines.

In Section 3.3 we discuss the *tobacco mosaic virus*. This was one of the first biological self-assembling systems to be understood. In 1955, H. Fraenkel-Conrat and R.C. Williams [43] demonstrated that a fully functional active virus could be self-assembled in a test tube from a simple collection of proteins and nucleic acids. While the structure of the tobacco mosaic virus is quite simple, the process by which it self-assembles is subtle and not obvious.

Finally, in Section 3.4 we examine the *ribosome*. The ribosome is one of the most important intracellular machines. Following the directions of messenger RNA it captures amino acids and builds proteins. That is, it receives information, processes that information, and uses that information to build physical structures. In short, it is a nanoscale production factory. The ribosome has also been shown to self-assemble. It can be reduced to a collection of 55 proteins and 3 RNA molecules. When placed in a test tube under the proper conditions, these components self-assemble into a fully functional nanoscale machine.

3.2 Proteins and Protein Folding

The cell, the basic unit of all living things, is a remarkably complex device. Cells have walls; a physical structure. These walls are active, full of channels opening and closing on the orders of tiny molecular gatekeepers. The walls are supported by a scaffold; the cytoskeleton. This scaffold not only supports the cell, but moves the cell. In turn, small machines walk along the cytoskeleton delivering packages and passing along information. A miniature postal network. The cell can sense light, heat, and pressure. The cell grows and divides. Sometimes cells group together forming large structures such as kidneys, livers, and you and I. But at the bottom, lies the cell. And at the bottom of the cell, serving both as the structural material and much of the machinery, are *proteins*.

Proteins are polymers. That is, they are long chain molecules composed of simple subunits. The subunits making up proteins are called *amino acids*. The basic structure of an amino acid is shown in Figure 3.1. Notice the presence of the *side chain group*. Variations in this side chain determine which amino acid we have. There are about twenty typical side chains that

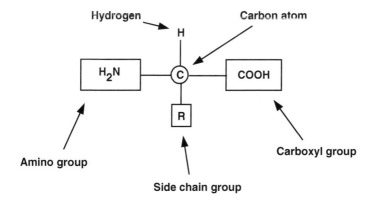

FIGURE 3.1: The basic structure of an amino acid. The side chain group, here denoted by R, can be one of about twenty different possible groups.

appear in biological systems, and hence twenty typical different amino acids that make up proteins. These amino acids are held together by *peptide bonds* and often a protein is referred to as a *polypeptide*.

When we discussed polymers in the previous chapter, we imagined them as being composed of individual, identical, monomers. This is the case for simple polymers like polyethylene, which is made up of individual ethylene subunits. But, for proteins, our alphabet is much larger. As we construct a protein, at each step we can choose from any one of twenty letters, the twenty amino acids. This means that the number of possible proteins, of even very short length, is enormous. For example, the set of small proteins one-hundred units in length has 20^{100} members.

Up until the early 1950's it was believed that the nature of an individual protein was completely determined by the amino acids of which it was constructed. That is, it was believed that if we simply knew how many of each type of amino acid made up a specific protein, we could completely characterize that protein.[1] In the early 1950's, the biochemist Frederick Sanger convinced the world otherwise by determining the amino acid sequence of the insulin molecule. Sanger developed techniques to find this sequence and demonstrated that proteins were not only specified by their amino acids, but by the order in which these amino acids were placed along the polypeptide chain. This order is now known as the *primary structure* of the protein. For this work, Sanger received the 1958 Noble Prize in Chemistry. He is one of only four individuals to have received two Noble Prizes. His second was awarded in 1980.

So, we have proteins, written in the amino acid alphabet, and we know that the order of the letters is important in determining the protein's properties. If proteins simply remained as long straight polypeptide chains, the story might be over. But, they don't. Proteins *fold*. The peptide bonds between the amino acids are strong, but flexible. We can imagine them acting like springs joining rigid spheres. The two spheres stay bound, but their relative positions can vary wildly. Perhaps the most remarkable fact in biology is that proteins fold, they fold quickly, and they fold *uniquely*. Determining the folded protein structure from the sequence of amino acids along its backbone is one of the central open questions in modern biology. This is the *protein folding problem*.

Typically, the folded structure of a protein may be decomposed into three parts. The *secondary structure* of a protein is specified by the sequence of α helices and β sheets that appear along its backbone. The α helix occurs when a protein chain assumes the typical helix motif present in much of biology. This is shown in Figure 3.2. In this structure, peptide bonds are attached to one another by hydrogen bonds. The β sheet occurs when a protein folds back and forth upon itself, like a roll of toilet paper allowed to unwind onto the bathroom floor.[2] In β sheets peptide bonds are again connected by hydrogen bonds. A typical β sheet is pictured in Figure 3.3. The *tertiary structure* of a protein specifies how the α helices and β sheets bend and pack into one another to form a basic protein subunit. Finally, individual protein subunits can bond

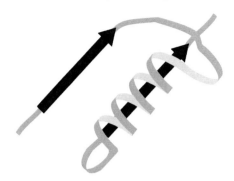

FIGURE 3.2:   The structure of an α helix.

together and form a complex protein molecule. This final structure is called the *quaternary structure*. This final structure is often also called the protein's *conformation*. Figure 3.4 shows the structure of a protein called CKD-2, or cyclindependent kinase. Several α helices are present. The complexity of the folded structure is also visible.

3.2.1   Understanding Protein Folding

In 1972, Christian B. Anfinsen received the Nobel Prize for Chemistry for his work on the protein folding problem. As he noted in his acceptance address [5], he was cited for

> ...studies on ribonuclease, in particular the relationship between the amino acid sequence and the biologically active conformation...

Unfortunately for us, Anfinsen did not quite solve the protein folding problem as presented above. He did however make the seminal contribution of showing that the recipe for the tertiary structure of proteins is contained in the chemistry of the amino acid sequence. By carefully severing the bonds that held the tertiary structure of ribonuclease together, he experimentally demonstrated that this tertiary structure would spontaneously reform and the protein would assume its original conformation. This key experiment led Anfinsen to propose the "thermodynamic hypothesis." This hypothesis says that the reason a protein assumes a particular conformation is because that conformation is thermodynamically the most favorable. That is, in a given environment, it migrates to the energetically most favorable state. He conjectured that the shape assumed by a protein is the global minimum of the free energy of the system. Anfinsen argued that the protein simply sampled the energy space, eventually winding up at the global minimum.

In contrast to Anfinsen, C. Levinthal argued [84] that the energy landscape accessible to a given protein was simply too large for the thermodynamic

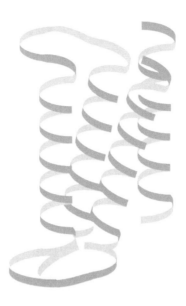

FIGURE 3.3:   The structure of a β sheet.

hypothesis to be valid. He noted that a small protein, 150 amino acids long, could assume any one of 10^{300} possible conformations. How could a protein sample such a large energy landscape and fold as reliably and as quickly as real proteins do? Instead, Levinthal offered a kinetic explanation. He argued that proteins must follow a well-defined *folding pathway*, determined by kinetics, and that a given protein must follow this pathway toward the global energy minimum.

After more than forty years of research, the debate still rages. The thermodynamic hypothesis is generally accepted, but modified by Levinthal's criticism. In the 1990's Onuchic and his coworkers [83] introduced the *folding funnel*, claiming that the energy landscape of a real protein has a deep funnel structure punctuated by shallow easily escapable local minima.

Whatever the outcome of the debate, which will certainly not be settled here, it is worth examining the basic idea behind protein folding models, both to give us a little more insight into the debate, and also a little more insight into how the subject of self-assembly can be approached. Perhaps the simplest class of models of protein folding are the *lattice models*. In this setup a protein is confined to live on a lattice like the one shown in Figure 3.5. Each circle represents an amino acid, or perhaps an α helix or β sheet. The dark lines joining the circles represent chemical bonds. The fact that the protein is confined to the lattice implies that circles must always lie inside the squares

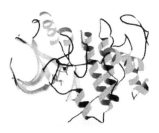

FIGURE 3.4:   The three dimensional structure of the protein known as CKD-2. Photo courtesy of the U.S. Department of Energy. Colors have been reversed for clarity.

as shown, and bonds must cross the lattice perpendicularly, not at corners. We may then formulate an energy for this protein. A typical approach is that of Shakhnovich et al. [119] who proposed an energy of the form

$$E = B_0 \sum_{ij}^{N} \triangle(r_i - r_j) + \sum_{ij}^{N} \triangle B_{ij}(r_i - r_j) \tag{3.1}$$

$$+D_2 \sum_{ij}^{N} \delta(r_i - r_j) + D_3 \sum_{ijk}^{N} \delta(r_i - r_j)\delta(r_i - r_k).$$

Here, r_i, gives the position of the ith circle along the chain, N is the total number of monomers, and $\triangle$ is the Kronecker delta assuming the value 1 if monomers are adjacent and zero otherwise. The parameter B_0 is assumed to be negative. Hence the first term on the right represents an attraction between nearby monomers. That is, adjacent monomers reduce the total energy. The second term allows the notion of the amino acid sequence to be built into the model. The B_{ij} represent the interaction energies between the ith and jth monomers. The fact that these can be different represents the different interaction possibilities between different types of amino acids. The last two terms introduce a penalty for monomers that attempt to occupy the same lattice site. Here, δ is the site Kronecker delta and assumes the value one if $r_i - r_j = 0$ and is zero otherwise.

Notice that Equation (3.2) is really an energy *functional*. We encountered the notion of a functional in Chapter 2 while studying bubble rafts. Here, the energy functional provides a map between the conformation of the protein and the energy. That is, given, in this case the vector specifying the shape of the protein, Equation (3.2) returns a real number identified as the energy of this particular conformation.

We can now view the problem in one of many ways. If we specify the B_{ij}, that is, if we specify the amino acid sequence of the chain, can we find the minimum energy conformation? This is one aspect of the protein folding problem. We could also imagine starting with a randomly shaped chain,

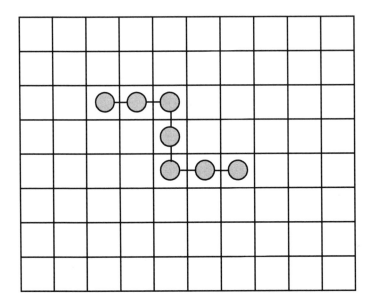

FIGURE 3.5: The lattice model of a protein. The circles represent amino acids. The dark lines represent the peptide bonds between amino acids. The protein is confined to the lattice.

disturbing the chain randomly, and allowing the chain to follow a downward energy gradient. This would let us examine the process of protein folding.

Whatever point of view we take, we are faced with the fact that there is no easy way to find the energy minima of Equation (3.2). If we allow the lattice to expand to three dimensions, or if we remove the lattice altogether, the situation only becomes worse. A tremendous amount of effort has been expended on this problem. The reader is referred to the review articles by Go [49] and Shakhnovich [120] as well as recent studies such as [52, 39, 40].

Try It Yourself - "Protein" Folding

As we've seen in this chapter, one of the central questions of modern biology is how the amino acid sequence of a protein determines its folded structure. In 2003, Sarah L. Keller of the University of Washington devised a simple, clever, experiment to explore how three dimensional structures could form by sequentially folding a rigid linear chain. She used the shape memory alloy *Nitinol* to construct simple models of proteins that could fold themselves. Her experiment is easily reproducible and allows one to develop an intuitive feel for the protein folding problem. To carry out this experiment you will need:

Materials

- Nitinol Wire. This is available from a variety of suppliers. See Appendix B for several sources.

- Heat Source. You'll need a heat source to anneal the wire. If you use low temperature wire, you will only need to heat the wire to 158°F. A hot water bath will be sufficient.

- Cold Source. A bucket of ice water works well.

- A rigid rod or wire to serve as the base for the nitinol shapes. Thick brass or copper wire is sufficient.

Procedure Construct a base model of the structure you wish to fold using the rigid brass or copper wire. Wind the nitinol around this structure to "train" it to assume the shape you want. Anneal this shape using your hot water bath or other heat source. Now, after the wire has cooled, remove it from the base, submerge it into the cold water bath, and straighten the wire. Now, slowly feed the straight wire into a hot water bath, allowing it to fold as you proceed. The Nitinol should assume your annealed shape.

Things to Try

- Explore the possible shapes that can be folded in this way. Can you coax the wire into a knot?

- What happens if the straight wire is plunged into the hot water bath? How does this folding problem vary from the linear sequential folding problem?

Further Reading The article by Sarah L. Keller contains many additional ideas for experiments with this system [68].

3.3   The Tobacco Mosaic Virus

At first glance the tobacco mosaic virus appears rather uninteresting. Electron micrographs such as Figure 3.6 show simple rod-like structures a few hundred nanometers in length and about ten nanometers in diameter. Not terribly exciting and certainly not as interesting as the complex structures of folded proteins. With a closer look the virus appears at least a little more interesting. The rod actually has a helical structure, a hollow core, and a strand of RNA bound inside the helix. If we counted, we'd find that the helical backbone consists of 2,130 identical protein subunits and that the RNA strand consists of 6,400 nucleotides.

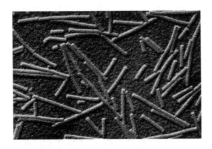

FIGURE 3.6:   Electron micrograph of the tobacco mosaic virus.   Credit: Omikron / Photo Researchers, Inc.

If we could watch the virus dancing about in solution outside of a cell, things still would not be very interesting. The tobacco mosaic virus is, after all, a virus. It exists in that weird borderland between the organic and inorganic. Outside of a cell it really does nothing. It has no internal metabolism, doesn't reproduce, and isn't very mobile. However, when it makes its way inside of a cell, it literally comes alive. If we could peer inside an infected cell, we'd see the protein subunits disassemble, the long strand of viral RNA work its way free, and the RNA hijack the enzymatic machinery of the cell turning the cell into a tobacco mosaic virus reproducing machine. From the point of view of the cell, this is of course bad, and this is one of the reasons this virus, and many others like it, are the subject of so much research.

But, there is one more feature of the tobacco mosaic virus that makes it of particular interest. The tobacco mosaic virus is known to self-assemble and the self-assembly process for this system is understood.

Self-assembly of the tobacco mosaic virus was first demonstrated by H. Fraenkel-Conrat and R.C. Williams [43] in 1955. They began with tobacco mosaic virus (TMV) particles isolated from infected tobacco plants. The

TMV was processed; this allowed them to break the virus into the individual proteins which constitute the helix, and into the nucleic acids which make up the encased RNA strand. Two different solutions, one containing protein, the other containing the nucleic acids, were prepared. All traces of TMV were removed from these solutions using standard centrifuge techniques. The two solutions were then mixed together and allowed to sit for at least twenty four hours. When the mixture was reexamined, Fraenkel-Conrat and Williams found fully assembled TMV. They had demonstrated that the tobacco mosaic virus could self-assemble in a test tube starting only with its basic protein and nucleic acid components.

But, how exactly does the self-assembly process for TMV proceed? The most obvious conjecture is that assembly of TMV is like crystal growth. Perhaps the RNA serves as a template for the proteins to start helix formation and then once this seed crystal has formed, additional proteins simply attach themselves to the growing helix, adding additional steps in a growing spiral staircase. But, the obvious answer is wrong. A large clue that something

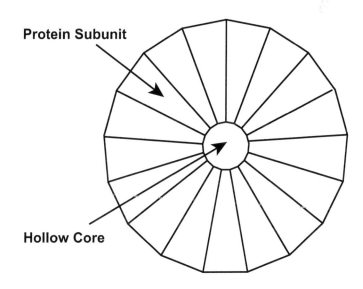

FIGURE 3.7:   Top view of the flat washer configuration of the tobacco mosaic virus proteins. There are 17 protein subunits in the washer. A complete washer is two layers deep and consists of 34 protein subunits.

more subtle is occurring lies in the original Fraenkel-Conrat and Williams experiments. Recall that in these experiments the assembly process took twenty four hours or more. This is fine in the nice sterile environment of a test tube, but in the wild, the unprotected RNA would be damaged beyond repair long before its protective protein coat was constructed. As pointed out by Butler and Klug [24] the solution to this problem lies within the protein subunits. In a system devoid of RNA, the proteins constituting the coat of TMV can self-assemble into a variety of forms. In a slightly alkaline solution the proteins exists as individual wedge-shaped subunits, or as small collections of these subunits. However, when the pH of the proteins' environment becomes neutral, a new structure appears. The proteins now assemble into the washer configuration of Figure 3.7. In this configuration, thirty four protein subunits assemble into two washers, one on top of the other. This neutral pH environment is similar to the conditions found inside a cell, hence most of the TMV proteins can be found in this form. When the environment is suddenly made more acidic (pH 5), something remarkable happens. These washers convert into a structure resembling an ordinary "lock" washer like the one shown in Figure 3.8. These lock washers can now stack and form a helix. Note that

FIGURE 3.8: Sequence illustrating the conformational change in the tobacco mosaic virus. On the left, the chain threads its way through a bound pair of washers just as RNA threads its way through the washer conformation of TMV. This causes the washers to flip into the lock washer state shown on the right. Photograph by the author.

this does not require the presence of RNA acting as a template, this structure is encoded in the protein itself. But, there is one remaining issue. We've seen how the proteins can form a helix, but this is devoid of RNA! How does the RNA work its way into the helical structure? And, what about the pH issue? We noted that for conditions like those inside of a cell, we don't get lock washers, we get ordinary washers. Well, the RNA-washer system neatly solves this problem. Given a strand of RNA and an ordinary washer, both of which can exist inside an ordinary cell under ordinary conditions, a loop of RNA inserts itself into a washer's hollow core and causes the washer to change its conformation and flip into the lock washer state. In turn, this conformational change traps a portion of the RNA inside the protein. The RNA dangling from this lock washer is then able to cause a similar conformational change for a second protein washer, this washer can then bind to the first, the process repeats, and a tobacco mosaic virus is born.

3.4   The Ribosome

In Section 3.2 we discussed proteins and the protein folding problem. We noted that proteins are both the building blocks and the machinery of the cell. In this section, we turn our attention to one these extraordinary machines, the *ribosome.*

The ribosome has been cited as the perfect example of a self-assembling nanoscale machine. In *Engines of Creation, The Coming Age of Nanotechnology* [34], the nano-pioneer Eric K. Drexler introduced the notion of an *assembler.* Essentially, an assembler is a nanomachine capable of building other nanomachines, including additional copies of itself. Drexler envisioned a device capable of grabbing atoms and molecules from solution, binding them together in a predefined way, and hence building structures with atomic precision. He imagined that these assemblers could be programmed and hence would be infinitely versatile, capable of producing any structure that did not violate basic physical law.

In many ways, the device envisioned by Drexler is present in all of us, in fact, is present in every living cell. That device is, of course, the ribosome. The ribosome is the cellular machine responsible for synthesizing proteins. This means that it must be capable of receiving genetic instructions, processing this information, and then using this information to build structures. This is not far from Drexler's assembler.

The basic operation of the ribosome may be easily understood. We refer to Figure 3.9. Note that the ribosome contains three binding sites labelled the A site, the P site, and the mRNA site. During the process of protein synthesis, a piece of mRNA attaches to the mRNA binding site. The "m" stands for "messenger." This is the structure that transports the information from DNA with regards to what the ribosome is to assemble. Once the mRNA is attached, a piece of tRNA binds to the P site on the ribosome. Here, the "t" stands for "transport." This piece of RNA is carrying an amino acid that the ribosome will use to build the desired protein. A second piece of tRNA then binds to the A site on the ribosome. The amino acids delivered by the tRNA are assembled according to the instructions of the mRNA still bound to

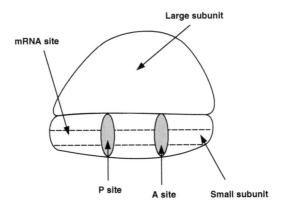

FIGURE 3.9: A highly idealized sketch of the ribosome. The A, P, and mRNA sites are used during the protein synthesis process.

the ribosome. The tRNA bound to the A site then moves a distance of three nucleotides along the mRNA. This ejects the piece of tRNA bound to the P site, drags the mRNA along like a tape in a tape recorder, and repositions the system so that the next amino acid may be bound. The reader interested in a more detailed description of this process is referred to [80] or [4].

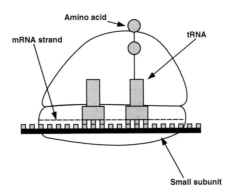

FIGURE 3.10: An idealized sketch of the ribosome during the protein synthesis process. Here, tRNA is bound to the A and P sites and a strip of mRNA is bound to the mRNA site.

So, the ribosome can synthesize proteins, and we know these proteins go on to fold and produce complicated three dimensional structures. What completes the notion of the ribosome as a perfect example of a nanomachine, is that it itself is built out of proteins and that these proteins can self-assemble in solution to form a fully functioning ribosome. The details of this process are described by Nomura in [96]. Roughly, we may describe the self-assembly of the ribosome as follows. From our schematic in Figure 3.9 we see that the ribosome is composed of two subunits. The detailed structure of the smaller unit is shown in Color Plate 11.7. The small subunit is composed of one RNA molecule and about 20 protein molecules. The large subunit is composed of two RNA molecules and and about 33 protein molecules. To successfully assemble a ribosome, these proteins and RNA molecules cannot simply be mixed in solution. Rather, the large and small subunits must be allowed to assemble separately. This can be done in a test tube under easily achievable laboratory conditions. Once the two subunits have assembled, in separate test tubes, these may be combined. The two subunits will now join and create a ribosome. Ribosomes created in this way have been shown to be fully functional.

3.5 Chapter Highlights

- *Proteins* are fundamental in biology. They are long chain molecules composed of amino acids. The sequence of amino acids along the polymer determines the final folded structure of the protein. How this sequence defines the final structure is one of the central questions of modern biology.

- The folding of proteins is a form of self-assembly. This assembly process may be modelled. Such models lead to large computational problems.

- The *tobacco mosaic virus* was one of the first biological systems shown to self-assembly. While it exhibits a classic helical structure, the process by which it assembles is subtle.

- The *ribosome* is an example of a self-assembled nanomachine. It serves as the manufacturing center for the cell, reading instructions delivered by RNA and producing specified proteins. It is comprised of protein subunits that self-assemble in solution and form a fully functional system.

Profile - John H. Reif

The familiar *barcode* appears on virtually every commercial product. Now, thanks to the efforts of a group led by *John H. Reif*, we even have nanoscale barcodes made entirely of DNA. But, you are not likely to find these barcodes gracing the side of your cereal box; Reif and his group designed their barcodes to illustrate the power of programmable self-assembly and perhaps someday serve as a display device for DNA computing.

Reif is an expert in both DNA computing and DNA based self-assembly. In fact, Reif exemplifies the interdisciplinary nature of the modern self-assembly researcher. His interdisciplinary interests go back a long while. While an undergraduate at Tufts University in the early 1970's, he eschewed the track of the traditional science major for their unique Unified Science Study Program. This allowed Reif to study an eclectic mix of topics spanning mathematics, computer science, physics, and engineering; an excellent background for a future self-assembly pioneer. As a graduate student he became necessarily more focused, earning both a Master's degree and a doctorate in applied mathematics from Harvard University. This broad approach to science has served him well. This is evidenced by wide range of his publications, spanning areas such as finance and investment theory, computational geometry, quantum computing, and robotics. But, perhaps his most interesting and important contributions have been to the fields of biomolecular computing and the self-assembly of DNA nanostructures.

In creating their DNA barcodes, Reif and his group made use of the hairpin loop structure of DNA. By attaching hairpin loops to a DNA double crossover molecule backbone, they could create a binary code. The presence of a loop in their structure represents a one, the absence a zero. By placing a seed particle in solution encoding a target binary string, the DNA barcode self-assembles, eventually reaching a size that can be imaged with an atomic force microscope.

In addition to DNA barcodes, Reif's group has made fundamental contributions to DNA tile based self-assembly and to the theory of algorithmic self-assembly. In 2003, Reif's group showed that DNA tile based self-assembly could be used to create *functional* nanostructures. Beginning with simple cross-shapes DNA tiles, the group made two dimensional strips and grids. The strips, when coated with metallic nanoparticles, could be used as nanoscale wires. The grids were decorated with a molecule that selectively binds to the protein streptavidin. This allowed them to produce a regular self-assembled protein lattice.

Presently, Reif is an Arthur Hollis Edens Professor in the Trinity College of Arts and Sciences at Duke University. He has been a Professor of Computer Science at Duke since 1986. He leads an active research group in DNA based self-assembly.

3.6   Exercises

Section 3.2

1. The number 20^{100} is so far beyond our everyday experience that it is difficult to get a sense of its magnitude. Compare this number to the estimated number of atoms in the universe. If we made each one of the possible 20^{100} one-hundred word proteins, stretched them out, and laid them end-to-end, how far would they stretch?

2. The literature on the protein folding problem is vast and has only been touched upon here. Search the literature and determine the state of the art in computationally solving the protein folding problem. How long does it take to computationally fold a protein? What is the longest protein that is computationally accessible? Compare the computational state of the art with nature.

3. Consider the lattice model of protein folding. Assume you have a protein consisting of only three amino acids. Write down all of the possible conformations of this protein and use the model of this chapter to compute the energy of each conformation. Now, show how to choose values of the parameters in the energy so that each conformation becomes the global energy minimum.

Section 3.3

4. Other biological structures are known to self-assemble. These include microtubules, bacterial flagella, and viruses other than TMV. Pick one of these and determine how it self-assembles and describe this process. How does it compare to the self-assembly of TMV?

Section 3.4

5. The function of the ribosome was described here rather abstractly. We did not touch upon the question of how the ribosome knows when to stop synthesizing a given protein. Investigate and explain how this works. Be sure to address the role of *stop codons*.

6. Describe precisely the steps and conditions under which the ribosome may be induced to self-assemble in the laboratory. See the article by Nomura [96] for these details.

3.7 Related Reading

The classic text on the molecular biology of the cell is the one by Alberts et al. The reader who wishes to delve more deeply into any of the topics discussed in this chapter is encouraged to begin with Alberts.

B. Alberts, D. Bray, J. Lewis, M. Raff, K. Roberts, and J.D. Watson, *Molecular Biology of the Cell*, Garland Publishing, 2002.

The book by Doi gives a nice introduction to lattice models in polymer physics.

M. Doi, *Introduction to Polymer Physics*, Oxford Science Publications, 2001.

The assembly of the tobacco mosaic virus and the story behind the quest to understand the process was recounted nicely by Butler and Klug in Scientific American.

P.J.G. Butler and A. Klug, *The Assembly of a Virus*, Scientific American, 242, 1978.

A readable introduction to the ribosome appeared in Scientific American.

J.A. Lake, *The Ribosome*, Scientific American, 245, 1981.

3.8 Notes

1. In retrospect this seems hopelessly naive. It is a bit like claiming the we could understand *Hamlet* if we simply knew all of the letters used by Shakespeare.

2. This is actually an interesting problem. Paper folding and coiling has been studied by Keller and Mahadevan. The same type of coiling has also been shown to occur in fluids. If you slowly squeeze shampoo onto a hard surface you can observe this fluid rope trick.

Chapter 4

Lessons from the Natural World

> *The progress of science consists in observing these interconnections and in showing with a patient ingenuity that the events of this evershifting world are but examples of a few general connections or relations called laws.*
>
> Whitehead, *Introduction to Mathematics*

4.1 Introduction

In this, the concluding chapter of *Part I: The Natural World*, we take a step back from the inorganic self-assembling systems of Chapter 2 and the organic self-assembling systems of Chapter 3 and attempt to abstract the general principles that nature uses to self-assemble structures. Abstracting principles from nature is a risky endeavor. It is certainly possible, perhaps even likely, that the schema we present here will be found lacking. It may happen that in the near or distant future other key principles used by nature in self-assembly will be uncovered. Or, that engineers will find methods to induce self-assembly that are completely unlike those in the natural world. Nonetheless, we make the attempt, if for no other reason than to encourage the reader to rethink the nature of self-assembly from an abstract viewpoint.[1] In Chapter 9, we'll return to this theme when we examine the theory of self-assembly. While it is possible to analyze the details of every different self-assembling system, the abstractions we begin to develop here will help unify and simplify our thinking later on.

At first glance, the systems we've studied thus far, crystals, ribosomes, bubble rafts, and viruses may seem to have little in common. But will we argue that when viewed from the vantage point of self-assembly, these systems are very similar. In fact, we'll argue that nature repeatedly uses the same motif in designing systems that self-assemble. This is true whether the systems are organic or inorganic, solid or liquid, atomic scale or galactic in size.

We'll begin in Section 4.2 by revisiting the first self-assembling system we encountered, the bubble raft. This time around, we focus less on the details of the self-assembly of the bubble raft, and more on the components that must be present in the system in order for self-assembly to occur. We'll see that there are four principle components: *structured particles*, a *binding force*, an *environment*, and a *driving force*. We'll describe how these four components come together and interact to produce the self-assembling bubble raft. Next, in Sections 4.2.1-4.2.4, we'll examine each of these four components in detail. We'll revisit all of the self-assembling systems considered thus far and identify the role that each of these components plays in each of these systems. We'll summarize the definition and role of each of these four components in Table 4.1. Finally, in Section 4.3, we'll examine other key components of nature's self-assembly motif. In this section we'll aim somewhere between the abstractions of Section 4.2 and the gritty details of the previous chapters. We'll examine three concepts that arise in the study of many different self-assembling systems. However, unlike structured particles, binding forces, an environment, and a driving force, the components discussed in Section 4.3 are not necessarily present in all self-assembling systems. We begin with the very important concept of energy minimization. This concept has certainly played an important role in many of the systems discussed thus far, but not all self-assembling systems are driven toward energy minima. This was alluded to in Chapter 1, and will be explored more fully in Chapter 7, when we examine self-assembling systems that remain structured only while dissipating energy. We also discuss the important concept of nucleation. This is arguably a feature of all self-assembling systems. Yet, in many systems, it is more a part of the process of self-assembly, and less a component of the initial design of a system. Finally, we examine template driven self-assembly. This again is an important concept, and as we'll see in *Part II: Engineered Systems*, has been used quite successfully in the design of self-assembling systems. But, it is certainly not universal in the sense that many self-assembling systems operate without the use of a template.

4.2   The Bubble Raft and Nature's Principles

In our discussion of bubble rafts, we learned how bubble rafts are made, why bubbles at the surface of a fluid are attracted to one another, why bubble rafts pack regularly, and even how bubble rafts can be used to demonstrate basic properties of crystals. But now, let's look at bubble rafts in a different light.

What exactly did we need in order to build a bubble raft? Or, more precisely, what features of the system had to be present in order to self-assemble a bubble raft? Well, obviously, we needed bubbles. But, instead of thinking of them as bubbles, let's think of them simply as *particles*. Next, we needed someplace for these particles to live, an *environment*. In the case of the bubble raft, the environment had several parts. These included the fluid on which the bubbles float, the tray in which the fluid resides, and the gravitational field in which the entire system is placed. Notice, each of these aspects of the environment is essential. Bubbles sitting on an otherwise dry table would not self-assemble, we need the fluid. The tray holds the fluid, but its shape also influences the size and shape of the assembled raft. Without gravity, there would be no buoyant force and hence no way to confine the bubbles to a surface. Other items might be present in a bubble raft environment, but these three are the crucial items. Now, the interaction between the bubbles, the fluid, and the gravitational field produced capillary forces, a *binding force* between bubbles. Without this interaction, there would be nothing to hold the raft together. This interaction also produced a *driving force* for the system. In moving together, the bubbles were reducing the gravitational energy of the system. In some sense, gravity was driving the bubbles together. If we abstract these four features from the bubble raft system, we can envision the system schematically as shown in Figure 4.1. These four features, *parti-*

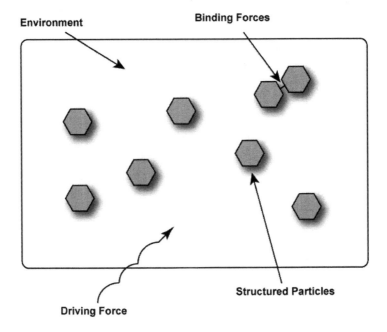

FIGURE 4.1: The basic features of a self-assembling system.

cles, an *environment*, a *binding force*, and a *driving force*, are present in all self-assembling systems. We note that sometimes the line between these four components is fuzzy. Binding forces blur into driving forces at large distances, particles respond and change in reaction to an environment, and it is often the environment that provides the driving force. Yet, it is still useful to think of these four as separate features of self-assembling systems.

4.2.1   Structured Particles

In the bubble raft, we had bubbles. These were the particles doing the self-assembling. In crystallization, we had atoms and molecules. In some sense, there is little difference between the bubbles forming a bubble raft and the sodium chloride molecules crystallizing into a chunk of salt. Both can simply be thought of as particles. When we get to polymerization however, we need to add the descriptive term *structured* to our simple notion of *particles*. Even in the elementary example of polyethylene studied in Chapter 2, we see that in addition to having distinct particles in a self-assembling system, these particles can have an internal structure that is crucial in the self-assembly process.

In the case of polyethylene, our individual particles are ethylene molecules. These are analogous to the bubbles of the bubble raft or the sodium chloride of salt crystals. Yet, unlike bubbles or sodium chloride, the ethylene molecules must change their internal structure in order to be able to bond to one another and create the polyethylene chain. Further, this change is not accomplished by the ethylene alone. In our simple model of polymerization, we introduced a second species denoted by A. The role of this second species was to induce a conformational change in the ethylene molecule by breaking one of its carbon double bonds, allowing it to bond to similarly transformed ethylene molecules. The A particles, which in a real system may themselves be very complex, acted as a catalyst for the polymerization reaction. Recall that the A molecules were used up in the reaction, sacrificing themselves in order to enable the creation of polyethylene.

In the case of micelles, the role of particle structure is even more dramatic. Here, lipid molecules are the building blocks of the micelle. They are the structured particles for this system. The detailed shape of the lipid molecule, the relative size of the head group and the tail, ultimately dictates whether the lipids will form a sphere, a cylinder, or another allowed structure. We could even imagine this situation occurring in a bubble raft if somehow we could make square bubbles. In *Part II: Engineered Systems*, we'll see systems essentially equivalent to a square-bubble bubble raft and the effect this has on the structures which form.

In organic self-assembling systems, structured particles are also the norm. We saw that proteins were long chain molecules composed of amino acids. If we think of the amino acids as being the particles in this system, then it is the choice of side chain that become crucial. That is, it is the side chain that adds structure to these particles. Any of the twenty or so biologically important amino acids could form peptide bonds with any other. But, according as their particular side chain, they interact with one another differently, inducing the protein to fold into α helices, β sheets, and eventually the more complex folded structures typical of proteins. We could also take another point of view.

We could imagine that the entire polypeptide chain constitutes the particle in this system. Then, the structure of the particle is clearly most of the story. We saw that the specific sequence of amino acids in this chain dictated how the protein folded. Hence, the ultimate structure of an individual protein molecule depended upon the initial unfolded structure of that very protein molecule. Once formed, individual protein molecules often come together to form a complex quaternary structure. Our canonical example of this was the ribosome, with two subunits formed from about 53 individual proteins. The shape of each of these proteins, that is the nature of these structured particles, is what allows them to bind in the specific manner needed in order to create a ribosome. Once again, we need individual particles, and their internal structure is crucial. Finally, the tobacco mosaic virus serves as another example of a system with structured particles. Recall that the individual proteins composing TMV naturally formed washer-like structures in conditions typically found in the cell. These washer-like structures adopted a different conformation when activated by RNA; they underwent a *conformational change*. In this new conformation they could bind and form the TMV helix. Again, the structure of the particle was crucial to the success of the system.

With these observations, a better schematic of a self-assembling system could be constructed. A schematic that allows for more structure and change in the nature of particles appears in Figure 4.2.

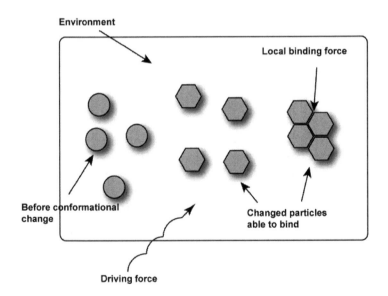

FIGURE 4.2:   A revised schematic of a self-assembling system.

Profile - Chad Mirkin

There are four winners of the Foresight Institute's Feynman Prize profiled in this book; Professor Chad A. Mirkin is one of the most recent of these recipients. Mirkin received the prize in 2002. He was cited for

> ...his work in opening up possibilities for the fabrication of molecular machine systems by selectively functionalizing nanoparticles and surfaces, particularly with DNA. This research enables the self-assembly of new structures, advancing the goal of molecular manufacturing.

Mirkin's work is not only useful, it is also often strikingly beautiful. One of the most dazzling images to grace the pages of *Science* in recent years was the Mirkin group's SEM's of self-assembled metal-polymer amphiphiles shown in Figure 4.3. In this remarkable study, the Mirkin group self-assembled micron sized spheres and cylinders from small gold-polymer rods. The rods behaved like amphiphiles, allowing the group to mimic the construction strategy used by nature in the building of micelles.

Mirkin was trained as a chemist. He received an undergraduate degree from Dickinson College and a doctorate from the Pennsylvania State University. Presently, Mirkin is the George B. Rathmann Professor of Chemistry, a Professor of Materials Science and Engineering, a Professor of Medicine, and Director of the Northwestern University International Institute for Nanotechnology. As a final measure of the impact of Mirkin's work, we note that he holds the distinction of being both one of the top ten cited chemist's and the number one cited nanomedicine researcher in the world.

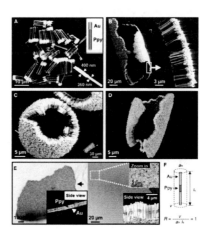

FIGURE 4.3: Self-assembled metal-polymer amphiphiles. From Park, et al., Science, v. 303, pp. 348-351, (2004), Reprinted with permission from the AAAS.

4.2.2 Binding Forces

The second component we identified in the bubble raft system was the capillary binding force. This is perhaps the most obvious of the four components; without *something* to hold particles together there would be no self-assembly. What is notable here is that we have already encountered a variety of possible binding forces. As we examine engineered systems, we'll encounter even more.

In the case of the bubble raft, it was the capillary force that served as the binding force for the system. In the case of crystals or polymers, it is a chemical bond. There are a wide variety of types of chemical bonds and a large menu of bond strengths from which to choose. In the case of a salt crystal, it is an *ionic bond* that joins the lattice together, the positively charged sodium ions being attracted to the negatively charged chloride ions. In polyethylene, we encountered *covalent bonds*. These strong carbon-carbon bonds occur as the result of the sharing of electrons. We saw that proteins were held together by *peptide bonds*. These occur when an amino group of one molecule reacts with a carboxyl group of another molecule. As we saw, these bonds are strong, but flexible.

Micelles presented us with a different type of binding force. Here, the binding force was not a result of pure interaction between individual lipid molecules, but reminiscent of the bubble raft, was a result of the interaction of lipid molecules with their environment. Lipids joined together in a particular configuration as their hydrophobic tails attempted to minimize contact with surrounding water molecules and their hydrophillic heads attempted to maximize this contact.

We mentioned above that the line between our four components was often blurry. This is especially true of the line between binding forces and driving forces in a system. After all, forces do cause changes in the motion of particles. It is hard to conceive of a force as not being a driving force. Typically, we distinguish the binding forces as those forces that hold a system in an ordered state.

4.2.3 Environment

That self-assembly needs an environment in which to occur is obvious, the effect of that environment on the process, perhaps less so. We've already discussed the environment for the bubble raft system. We saw that this environment consisted of three components, the fluid, the container, and the gravitational field, and that each of these three components was necessary and affected the assembly process. In the exercises for this chapter, we ask the reader to ponder this further by envisioning a bubble raft experiment where the strength of the gravitational field can be varied.

In the other systems we've encountered so far, the environment plays just as crucial of a role. Crystals typically precipitate out of a solution. This solution serves as the environment for the crystal system. Both the temperature and concentration of this solution, this environment, affect the product that results. If the system is cooled very rapidly, crystals will precipitate rapidly and we'll obtain dendritic structures like those in Color Plate 11.6. If the system is kept warm, or is very dilute, the process will proceed more slowly and we'll obtain larger, more uniform structures like those in Figure 2.11.

In the case of polymers, or micelles, the environment is also typically a solution. Changes to environmental variables such as temperature and concentration once again effect the end product. In the case of micelles, we encountered the critical micelle concentration. That is, it was not until the environment reached the proper conditions that micelle formation could even occur. Another factor affecting the formation and type of micelles that form is the salinity of the solution. Just as changes in the relative sizes of the head group and tail of lipids could induce the micelles to form different structures, changes in the salinity of the surrounding fluid can induce similar changes. The reader is also asked to explore this further in the exercises.

The environment also plays a key role in the self-assembly of biological systems. Protein folding and the assembly of the ribosome occur in conditions particular to the living cell. In order to be replicated in the laboratory, such conditions must be replicated. The effect of the environment in our biological examples was clearest in the case of the tobacco mosaic virus. We saw that the assembly of TMV protein subunits was controlled by the pH of the solution in which they were immersed, an environmental variable. If the solution was alkaline, the proteins only formed small clusters of protein subunits. As the pH was adjusted and the solution became neutral, the proteins formed washer shaped complexes with a precisely defined height and number of subunits. As the pH was increased and the solution became more acidic, these washers changed conformation and flipped into the lock washer state.

4.2.4   Driving Forces

The final essential component of self-assembly was the driving force. In our schematic of a self-assembling system, Figure 4.1, we envisioned the system as a collection of randomly placed particles floating in an environment. Once these particles have reached a particular state, a binding force captures the particles, orients them properly, and holds them there. But, it is the *driving force* that causes the system to move through and sample different possible configurations of the system on its way to the ordered final state.

In crystallization, the driving force is thermal noise. The random motion of molecules in solution force the sodium and chloride ions to dance and jiggle about until they randomly come together at an appropriate binding site on a growing crystal lattice. Here, our bubble wrap ping-pong ball analogy breaks down. It is not that we are dropping ping-pong balls onto a sheet of bubble wrap, rather it is that our living room is filled with active ping-pong balls bouncing around due to thermal noise. Occasionally, a ping-pong ball hits the bubble wrap at a void and sticks.

Polymers also grow in solution and the process is driven by thermal noise. The ethylene molecules are not ordered in some microscopic assembly line, awaiting their date with a catalyst and the opportunity to bind to a growing chain. Rather, again, the situation is messy and random. Ethylene molecules bounce about in solution, driven by thermal noise, occasionally encountering a catalyst and interacting appropriately.

For micelles, proteins, ribosomes, and the tobacco mosaic virus, the situation is similar. The image we should have in mind is a noisy system where particles hop, shake, and shimmy at random until they've arrived in a position where a typically short range binding force can drag them into some final ordered state.

4.3 Other Aspects of Nature's Motif

Structured particles, binding forces, an environment, and driving forces. These all clearly play a role in nature's design of self-assembling systems. But, surely this cannot be the complete story. And, it's not. If we think back to the bubble raft, it can indeed be thought of abstractly like the schematic in Figure 4.1. Yet, we could not have performed any of the analysis of Section 2.2.2 purely from the schematic of Figure 4.1. In order to compute the shape of the meniscus between two bubbles, we not only needed to know that there was an environment, but we needed to know how that environment responded and changed in the presence of a bubble. If we want to understand how a specific protein folds, it is not enough to know that it is comprised of twenty different amino acids, but we must also know how those amino acids interact with one another. The devil is in the details and in these details we'll find many other facets of nature's self-assembly motif.

Viewed this way, there are at least as many facets of self-assembly as there are different self-assembling systems. But, just as it was useful in Section 4.2 to begin to think of self-assembly from an abstract viewpoint, it is useful here to consider a few of the major facets that reappear time and time again.

4.3.1 Energy Minimization

It is tempting to think that all self-assembling systems are driven by some principle of energy minimization. One is almost willing to posit a "thermodynamic hypothesis" of self-assembly and claim that like proteins, all self-assembling systems sample some state space, eventually making their way to a local or global energy minima. Unfortunately, such a hypothesis would be wrong. As we'll see in Chapter 7, there are in fact many systems that remain ordered only while dissipating energy. Recall that in Chapter 1, we defined

TABLE 4.1:   Nature's Four Key Components of a Self-Assembling System

Structured Particles	These are the components that actually do the assembling. The inherent structure of the particles determines the complexity of the final self-assembled system. Tailoring the structure of the particles provides the first means of control of the self-assembling system. Often, the particles have an internal structure that can be changed by external stimuli. This internal structure is often called the *conformation* of the particle. Examples of structured particles include magnets, lipids, and proteins.
Binding Forces	This is what holds the particles together. Usually, this binding is reversible, allowing the system to move from local equilibria to a global equilibrium. Altering the binding force provides a second means of control of a self-assembling system. The interaction of the particles or external stimuli may activate or alter binding forces in the system. Examples of binding forces include capillary forces, electromagnetic forces, and chemical bonds.
Environment	The particles are embedded in an environment. The proper environment is necessary in order for the binding force to act. For example, capillary forces are only useful when the particles live at the surface of a liquid. Tailoring of the environment or dynamically changing the environment provides an important means of control of a self-assembling system. The environment can be manipulated to change the conformation of particles or to alter binding forces in the system.
Driving Force	In order for self-assembly to occur the particles must interact stochastically. This driving force in the system is usually thought of as *noise*. This may be thermal noise, physical oscillation of the system, or driving via electromagnetic fields.

these non-equilibrium self-assembling systems as *dynamic* self-assembling systems.

Perhaps an example is in order. When we introduced the diffusion limited aggregation model (DLA) of crystal growth in Chapter 2, we noted that DLA had been applied to the study of bacterial colonies. But, there is an important difference between a bacterial colony and a crystal. The bacterial colony is alive, it is constantly moving and evolving, and it is using energy. If the food source is removed from the colony, it collapses, dies, and the structure is lost. The non-equilibrium structure of the bacterial colony persists only so long as it is dissipating energy. It is most decidedly not pushed towards or operating at a global or local energy minimum.

However, nature often does act to minimize energy in a system and this does often control the process of self-assembly. We saw the principle of energy minimization at work when we computed the shape of a meniscus near a rigid wall. Nature chose the shape that minimized the total energy, in this case, the sum of the gravitational potential energy and the surface energy of the fluid. In the bubble raft, the principle of energy minimization provided an attractive force between bubbles. The system pushed bubbles together in an attempt to minimize an expenditure of energy. Many of the other systems we've examined could be viewed in the same way. Crystals are minimizing a chemical potential, lipids are minimizing an interaction energy with water molecules, and proteins are folding as the system slides down an energy funnel. When examining or designing self-assembling systems, we must certainly bear in mind nature's tendency towards economy.

4.3.2 Nucleation

The notion of nucleation may not seem on par with the grand principle of energy minimization, and perhaps it's not. But, nucleation is a common feature of self-assembling systems and deserves at least a brief mention.

Nucleation is easiest to understand in the context of crystal growth. In our discussion of crystal growth, we imagined starting with a large bubble wrap crystal surface arranged nicely on our crystal growing floor. But, how did that bubble wrap appear in the first place? If we start with a solution, a warm soup of molecules bouncing about, how do we get to the frozen structure of a crystal? This requires nucleation. Typically, we think of nucleation as a localized phase change. In crystallization, this is a rare event and we usually need a seed crystal, that is a crystal that is already formed, to get the growth process started. Other forms of nucleation include the use of a catalyst. When we considered polymerization, in particular of polyethylene, we saw that a catalyst was necessary in order to break a carbon-carbon bond and allow the monomers to form a chain. While in principle this could occur without the presence of a catalyst, it is such a rare event that it is not observed to occur.

In self-assembly, both the concept of pure nucleation and the concept of a seed particle play a role. In an ideal self-assembling system, such as in Figure

4.1, we imagine starting with a collection of identical particles and simply allowing the process to proceed without external interference. But, this implies that self-assembly must start *someplace*. If the particles carry information, such as may be encoded in their conformations, the initial nucleation event may lead to a specific, and perhaps undesirable, assembly pathway. A way to overcome this is to start with a seed particle. This guides the system down a chosen path towards a desired end. Both of these concepts will be explored more fully in Chapter 9.

4.3.3 Templates

Templates are commonly used in the fabrication of macroscale objects. We pour cupcake batter into cylindrical tins, chocolate into flat molds, and liquid metal into specially designed dies to form coins. All of these are examples of templates. Nature makes use of templates as well. When we examined the ribosome in Chapter 3, we saw how the ribosome built proteins by capturing mRNA and using it as a pattern to construct the appropriate protein. In turn, the RNA was built in a different template driven process where DNA served as the template.

In the context of self-assembly, a template can be used to align or orient particles so that they may bind via a secondary process or so that binding occurs more rapidly. Schematically, we can envision template driven self-assembly as is shown in Figure 4.4. Here, the environment includes a template. Rather than binding to one another, the particles first bind to the template. The particles in the template then become bound together, creating the final structure. We will explore several examples of engineered systems that use templates in *Part II: Engineered Systems*.

4.4 Chapter Highlights

- Nature's self-assembling systems have four key components in common. These are *structured particles*, a *binding force*, an *environment*, and a *driving force*.

- The structure of particles often dictates the final self-assembled structures in a system. Particles can also change their *conformation* during the self-assembly process.

- There are a wide range of binding forces to choose from in self-assembly. Chemical bonds provide an especially full menu of forces from which to choose.

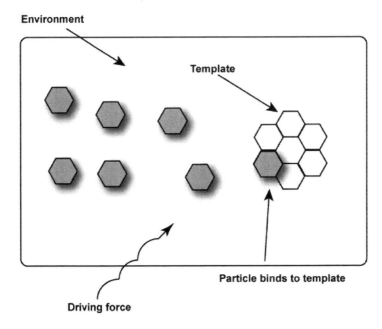

FIGURE 4.4:   A schematic of a template driven self-assembling system.

- The environment plays a crucial role in any self-assembling system. Interactions between particles and the environment often control the conformation adopted by a particular particle and hence the shape of the final self-assembled structures.

- The driving force is the stochastic or random component of self-assembly. The wonder of self-assembly is that a system driven by noise leads to remarkably ordered structures.

- Other features of nature's design of self-assembling systems include the principle of energy minimization, the process of nucleation, and the use of a template.

4.5   Exercises

Section 4.1

1. In *Part I: The Natural World* we discussed three organic and four inorganic naturally occurring self-assembling systems. Find other examples of each and discuss how they fit into the framework of Table 4.1.

2. In Chapter 1, we offered a definition of self-assembly. Explain how this definition is related to the principles of Table 4.1 and why each statement of the definition is necessary.

3. Reconsider the bubble raft of Chapter 2. If the bubble raft system were moved to a gravity free environment, would the bubble raft self-assemble? Suppose the gravitational force could be slowly decreased. What would happen to the bubble raft system?

4. Reconsider the self-assembling micelles of Chapter 2. In our discussion, we focused on the concentration of lipids in a solution and the relative size of the head group compared to the tail. However, the *salinity* of the solution in which the lipids are immersed also affects the shape of the structures formed by the micelles. How does this occur? Which aspects of the four key components are affected by changes in salinity?

5. There are many types of chemical bonds that occur in nature. We've discussed ionic bonding and covalent bonding. What other types of bonds are possible? How do the strengths of these various bond types compare?

4.6 Notes

1. We make no claim to originality in this chapter. The ideas presented here are gleaned and distilled from a careful reading of the increasingly vast literature on self-assembly. We encourage the reader to delve into the literature and rethink the basics of self-assembly for themselves.

Part II

Engineered Systems

Chapter 5

The "Cheerios Effect" and Other Simple Systems

If two fresh Cheerios are placed near each other while floating on milk, they will rapidly pull together. What force causes that attraction?

Jearl Walker - *The Flying Circus of Physics*

5.1 Introduction

In Part One: *The Natural World*, we explored naturally occurring organic and inorganic systems that exhibit self-assembly. We saw that nature utilizes four key principles in her design of such systems: *structured particles, binding forces, environment, and driving forces*. These are listed and explained for reference in Table 4.1.

In this, the second part of the text: *Engineered Systems*, we turn our attention to man-made self-assembling systems. Our goal is to answer the question: "What self-assembling systems can we build now and how do we do it?" In this and the next three chapters we will examine this question from the viewpoint of physics, chemistry, biology, and engineering. We will see that no matter what viewpoint we take, the four basic components of nature's design are utilized. We will also see that from an engineering perspective, each of these principles provides an opportunity for design.

However, we'll also find that the design of self-assembling systems presents the designer with three major challenges. We call these the *forward problem*, the *backward problem*, and the *yield problem*. The forward problem is predictive. Given a set of particles, a binding force, an environment, and a driving force, what structures will the system produce? The backward problem[1] goes the other way around. Given the desired structure, how do we choose a set of particles, a binding force, an environment, and a driving force to assemble this structure? The yield problem arises because of the stochastic nature of the driving force in self-assembly. Typical systems contain local energy minima and our actual output is different from our desired output. The yield problem asks: How do we maximize our yield?

In this chapter we explore three very simple types of self-assembling systems that illustrate how natures principles are present in man-made systems and illustrate the forward, backward, and yield problems faced by the designer. First, we examine a purely mechanical system due to Penrose and Penrose [100]. This highly simplified system will show how nature's key principles come into play in an engineered system. Next, we will examine simple magnetic self-assembling systems. These will illustrate the forward, backward, and yield problems. Through these examples we will introduce the notion of complexity in self-assembly and offer a preliminary definition and way to measure complexity. As our third "engineered" self-assembling system we will consider the "Cheerios Effect." This system is related to the bubble raft of Chapter 2, but here, bubbles are replaced by floating solid particles. In addition to describing this system and discussing how nature's principles come into play, we will also develop the theoretical machinery necessary to understand this system in detail. This system, although seemingly frivolous, serves as an excellent demonstration of the mechanisms behind several serious and practical self-assembling systems discussed in the next chapter.

5.2   The Penrose Model

When we examined natural self-assembling systems in Part I of this text, the complexity, especially that of the organic systems, shone through. Because of the complexity found in nature, we might assume that an engineered self-assembling system need be similarly complex. Remarkably, in 1957, Penrose and Penrose [100] showed that a very ordinary system, made from very ordinary materials such as plywood or plastic, can exhibit self-assembly. The basic design of their system is shown in Figure 5.1. Their system consists

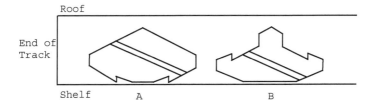

FIGURE 5.1:   The self-assembling or self-replicating system of Penrose and Penrose. Note that there are two types of particles, A and B, and in this figure they are unassembled.

of two types of particles, labelled A and B. The particles are flat, roughly two-dimensional objects, and can be cut out of plywood or even stiff cardboard. The important feature of the particles is their shape. As can be seen in Figures 5.2 and 5.3 they are cut in such a way that when properly oriented they can interlock. These particles are then placed on a track. This track

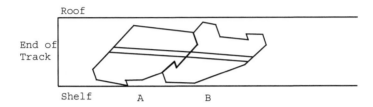

FIGURE 5.2: The self-assembling or self-replicating system of Penrose and Penrose. In this figure elements A and B are hooked together.

restricts their motion; while they are able to move up and down slightly and change orientation to some degree, particles cannot pass one another on the track. If a number of particles of types A and B are placed along the track

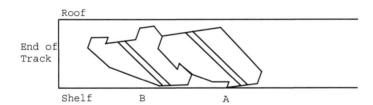

FIGURE 5.3: The self-assembling or self-replicating system of Penrose and Penrose. This figure shows an alternative complex formed by particles A and B.

in an unhooked state and the track is shaken *horizontally*, the pieces will not link up. If however, an A-B complex, as shown in Figure 5.2, is added to the system and the track is again shaken horizontally, something truly remarkable occurs. Namely, if an A particle is immediately to the left of a B particle and immediately to the right of the A-B complex, the complex will reproduce itself, causing a new A-B complex to form. A close examination of Figure 5.2 reveals why. Under random agitation, the A particle to the right of the

TABLE 5.1:   The Effect of a Seed Particle on the Penrose Model

Initial String	A B B A B A
Result after adding an AB complex	**AB** AB B AB A
Result after adding a BA complex	**BA** A B BA BA

complex slips under the right side of the complex. This causes a large rotation of the A particle and puts it into position to link with the B. A similar situation occurs if the alternative complex, B-A, shown in Figure 5.3 is used. An example of the effect of an A-B or B-A complex on an initial string is shown in Table 5.1.

5.2.1   Nature's Principles in Action

The Penrose Model is often cited by self-assembly researchers as an example of a self-assembling system and by self-reproduction researchers as an example of a self-reproducing system [44]. While the main interest of Penrose and Penrose was in self-reproduction, their model still serves as a good example of a self-assembling system. In contrast to many of the engineered self-assembling systems we will examine in the next few chapters however, the Penrose Model requires the presence of a *seed*. That is, if no initial complexes are present in the system, no new complexes will be formed. Once a seed is injected into the system, it catalyzes the formation of other identical complexes. This is reminiscent of the use of a seed crystal to start crystal growth in a natural system or the use of a catalyst to start a polymerization reaction in a polymer system.

Here however, our interest is in examining how the Penrose Model makes use of the four self-assembly features in Table 4.1. First consider *structured particles*. Clearly, the particles in the Penrose Model have a built-in structure. They are designed first, so that mechanical binding is possible, and second, so that an interaction leading to binding requires the presence of a seed complex. Also, the system uses precisely two types of particles and these particles do not undergo any conformational changes. The *binding force* in the Penrose Model is purely mechanical. The particles "lock" together and stay locked under mechanical forces. The *environment* and *driving force* in the Penrose Model are particularly interesting and the model well illustrates the interaction of the two in a self-assembling system. Penrose and Penrose were very careful to specify that the track be shaken *horizontally*. This effectively restricts the motion of the particles to one dimension. Although some small vertical motion is unavoidable in practice due to particle collision, the motion is generally too small to place the particles in an orientation where they can bind without the presence of a seed particle. This is an essential feature of the Penrose Model when viewed from the perspective of self-reproduction. If we allow random shaking of the system, some A-B and B-A complexes will form *without* the presence of an initial seed. However, in this system the mechanical binding force is weak and these complexes will be destroyed soon after they are created.

The basic system developed by Penrose and Penrose has been extended and exploited by others. An interesting article on simple mechanical and electromechanical self-assembling systems was presented at a Foresight Conference by Lohn et. al. [86]. In their paper they discuss the Penrose Model as well as another early mechanical model of self-assembly that used railroad cars. They also present a plan for two electromechanical self-assembling systems. To date, it appears that neither of these have actually been built. Details of these systems and other macroscale self replicating machines may be found in [44].

5.3　Magnetic Self-Assembling Systems

Take a stack of disk-shaped magnets, lay them out randomly in a container, and give the container a shake. We can all solve the *forward problem* for this simple experiment; the magnets jump together and form a chain or cylinder. While very simple, this self assembling system again contains Nature's four self-assembly components of Table 4.1. The magnets are the *structured particles*, magnetism is the *binding force*, the container forms the *environment*, and our shaking is the *driving force*. However, unlike the Penrose Model, magnetic self-assembling systems promise more than theoretical insight. As we'll see in this section, the basic idea of shaking magnets in a container has already been extended to practical engineering applications.

But first, we note that thinking about magnets in a container yields further insight into the process of self-assembly. The first observation to make about the magnet-container experiment is the low complexity of the self-assembled structure. Basically, we input a bunch of dipoles and the output is a dipole chain. The order of the magnets in the chain is purely random, as is their angular orientation with respect to the axis of the chain. In fact, it is hard to imagine a smaller gap in complexity between an initially disordered system and the final assembled structure. This however raises several questions: What do we mean by complexity? How do we define complexity? How do we measure complexity? If our goal is to induce systems to self-assemble into "complex" structures these are questions we'll need to address. By the end of this section, we'll offer a definition of complexity and a way to measure the amount of complexity in a system. We'll again return to the notion of complexity in Part III of this book when we more fully examine mathematical models of self-assembly.

There is still more to be learned from the simple magnet-container system. In a simple variant of this experiment Sharpe [121] embedded tiny disk shaped magnets into 1cm × 1cm plastic cubes. Embedding only two magnets in each cube, one in each of two opposing faces, a simple alphabet guaranteed to assemble into linear chains was constructed. Again, we should think of this in terms of the *forward problem*. Given the design of the Sharpe experiment, the assembled structures may be predicted. Note that in designing the experiment, the *backward problem* also needed to be considered. The goal of the Sharpe experiment was to form only linear chains. The cubes, the magnets, their arrangement, and the shaking of the system were all chosen to guarantee that the system produced only linear chains; i.e., the backward problem was solved.

If we view the initial magnet-container system as a collection of identical dipoles, we see that the magnet-cube system can contain dipoles, monopoles, and "end caps." If magnets are embedded into a given cube such that one north face is outward and one south face is outward, the cube is a dipole. Embedding the magnets such that two north or two south faces are outward produces a monopole. Finally, embedding only one magnet, with either a north or south face pointing outward, produces an end cap. In the experiments of Sharpe, the cubes were arranged on a flat surface bounded by a circular wall and the system was vibrated vertically causing the cubes to bounce and interact. Vibrations were kept small to ensure that the cubes did not move in three dimensions but only interacted in a plane. The cubes and their initial layout on the shaker surface are shown in Figure 5.4.

FIGURE 5.4: The cubes and their initial layout as used in the Sharpe experiments. Credit: K. Sharpe/MEC Lab - University of Delaware.

The Sharpe experiments clearly illustrate two features of engineered self-assembling systems that are not obvious when considering the system from a purely theoretical viewpoint. As discussed above, our intuition about a small collection of magnets leads us to expect that the system will form one long chain. Yet, when the system becomes larger, or constrained, as in the Sharpe experiments, the *yield problem* of self assembly becomes apparent. In our magnet-container experiment we imagined that our goal was to assemble a chain. Perhaps we wished to assemble a single chain of maximal length, or perhaps we wished to assemble multiple copies of shorter chains. Unfortunately, our interactions are stochastic, so we never get quite what we want. In the Sharpe experiments, the system was vibrated until an apparent equilibrium was reached. At this point the number of chains of different lengths was counted yielding a distribution curve such as the one shown in Figure 5.5. Note that relatively few long chains were produced, that roughly thirty

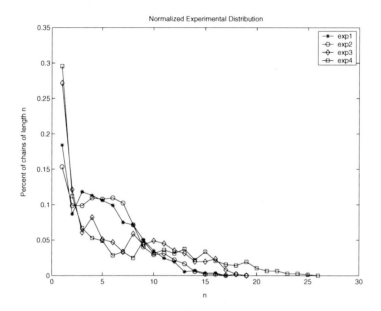

FIGURE 5.5:   Normalized distribution of chains formed in the Sharpe experiments. Credit: K. Sharpe/MEC Lab - University of Delaware.

percent of the chains contained only two cubes, and that while the system started with 57 cubes, no chain longer than 25 cubes was produced.

In observing video of the cube experiment[2], Sharpe noted a second feature of engineered self-assembling systems; the yield depends not only on the structured particles, but on the details of the interaction between the particles, the environment, and the driving force. If the system is driven too hard, no chains form; mechanical collisions lead to forces that dominate the binding force. If the system is driven too softly, the time to form any chains becomes exceedingly long. More importantly, as the system evolves, the way that the particles interact changes. As long chains were created in the cube experiment they began to form barriers to the motion of shorter chains and individual cubes. The size of the container sets the maximal chain length, and in turn the presence of maximal chains sets the length of shorter chains.

The yield problem and the interaction of the particles with the environment and the driving force, all present challenges to the designer of self-assembling systems. Even this simple magnet-container system, which at best leads to low complexity structures, presents us with these difficulties. From an engineering perspective, overcoming these difficulties is crucial if self-assembly is to be made practical. As an extreme example, imagine our goal was to assemble thousands of magnet chains consisting of precisely ten magnets each. Imagine our strategy was to toss all of our magnets into a large container and agitate the container. Now, our intuition about the behavior of a handful of magnets in a container no longer serves us well. As in the Sharpe experiments, when we open our gigantic container, we can expect a gigantic mess, with relatively few usable length ten magnet chains.

5.3.1   Pattern Formation in Magnetic Spheres

In a different sort of magnetic self-assembly experiment, the group led by Wolfgang Losert of the University of Maryland [128, 129] studied the interaction of magnetic spheres. Their work highlights the role that structured particles play in the self-assembly process and illustrates how an energy analysis helps the designer understand the behavior of the system. In the Losert experiments, cylindrical magnets were encased in plastic spheres. The spheres were colored so that the north pole of the magnet corresponded to a light hemisphere and the south pole to a dark hemisphere. This allowed dipole orientations to be observed. In contrast to the Sharpe experiments, where the cubical structure of the particles restricted the system to straight chains, the spheres of the Losert experiments allowed for the dipole orientation to change slowly throughout the final assembled structure.

The basic experimental setup can be seen in Figure 5.6. In the Losert experiments, cylindrical magnets of a given length, s, and diameter, D, were used. The long magnets had length $s = 1.42$cm and diameter $D = 0.94$cm, while the short magnets had length $s - 0.64$cm and diameter $D = 0.95$cm. When encased in the hard plastic spheres both types of magnet's spheres had diameter $d = 1.69$cm. The magnets were randomly poured into a cylindrical container with a diameter 17.5 times the particle diameter. The height of the container was 1.7 times the particle diameter. The top and bottom of the containers were sealed, but the top was clear so that images could be captured. They then vertically vibrated the container and observed the effect.

The outcome of the experiments with the short magnets can be seen in Figure 5.6. Note, that these experiments reveal a dependence of the final configuration on initial particle density. That is, the self-assembled structure depends strongly on the environment. In Figure 5.6 we see that an initial low-density configuration leads to the assembly of a simple chain-like network. As the density is increased, a large macrovortex configuration is achieved. These two patterns appeared starting from an initial random configuration of the particles. Stambaugh et al. also conducted experiments with an initially or-

FIGURE 5.6:   Magnetic spheres exhibiting pattern formation. (a) shows an intermediate density system (95 spheres) shaken at 4.3g for 250s. (b) shows a dense system (190 spheres) shaken at 4.3g and allowed to reach a steady state. (c) shows the microvortex pattern made by square-packed setup, undriven spheres. (d) shows a hollow macrovortex state of spheres after being set up in an aligned, concentric rings pattern and shaken at 4.3g. Reprinted with permission from Stambaugh et al. , Physical Review E, v. 68, 2003, 026207. Copyright 2003 by the American Physical Society.

dered configuration of particles. As can be seen in Figure 5.6, when a square packed initial setup was used, a microvortex configuration was achieved. Finally, when a initial configuration consisting of aligned concentric rings was used, the final state was a hollow macrovortex.

Switching to randomly distributed long magnets, a different result is seen. In Figure 5.7, we see that a dense initially random state becomes a concentric ring state after shaking. When the shape of the container is changed to one with a square cross-section, the same basic final state is achieved, this time with squared off ends.

While the outcome of the experiments clearly shows a dependence on initial density and particle type, the reason is not necessarily clear. To explain these

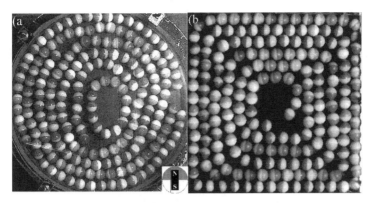

FIGURE 5.7:   More magnetic spheres exhibiting pattern formation. (a) shows the final state of a dense system after being shaken at 7.8g for 250s. (b) shows the final state after being shaken in a square box at 7.8g for 250s. Reprinted with permission from Stambaugh et al., Physical Review E, v. 68, 2003, 026207. Copyright 2003 by the American Physical Society.

observations, the Losert group considered the energy of the system. They modelled each of their particles as a sphere of diameter d, containing two separate charges $\pm q$, separated by a distance s. The product, qs, was kept constant to model a particle with constant dipole moment. The energy of a given configuration was then calculated from the standard Coulomb energy formula:

$$U_{ij} = \frac{\mu_0}{4\pi} \frac{q_i q_j}{R_{ij}}. \tag{5.1}$$

Here, μ_0 is the permeability of free space, q_i is the ith charge, and R_{ij} is the distance between the ith and jth charges. Summing over all interparticle charge pairs yields the total energy of a given system, U.

The experimental results of the Losert team may be understood by considering the energy of a pair of contacting particles. While these results may be found in [128] it is instructive to repeat them here. In Figure 5.8 we show the coordinate system we'll use for these calculations. In this system, we have four interacting interparticle charges. Hence, the total energy may be written as

$$U = U_{13} + U_{14} + U_{23} + U_{24}. \tag{5.2}$$

The coordinates and charges of the four particles are shown in Table 5.2. From this, the distance between charge pairs is easily computed. We find:

$$R_{13} = \sqrt{\left(d - \frac{s}{2}\cos(\theta) - s\right)^2 + \frac{s^2}{4}\sin^2(\theta)} \tag{5.3}$$

$$R_{14} = \sqrt{\left(d + \frac{s}{2}\cos(\theta) - s\right)^2 + \frac{s^2}{4}\sin^2(\theta)} \tag{5.4}$$

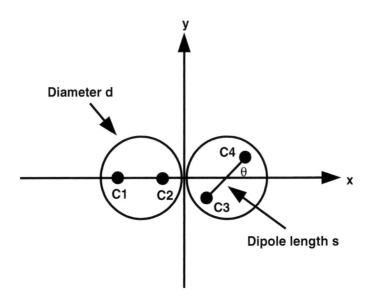

FIGURE 5.8: The coordinate system of our two dipole system. The charges and coordinates of the four points are shown in Table 5.2

TABLE 5.2: The coordinates and charges for calculating the energy of our two dipole system.

Point	Coordinates	Charge
C1	$(s - d/2, 0)$	$-q$
C2	$(s/2 - d/2, 0)$	q
C3	$(d/2 - s/2\cos(\theta), -s/2\sin(\theta))$	$-q$
C4	$(d/2 + s/2\cos(\theta), s/2\sin(\theta))$	q

$$R_{23} = \sqrt{\left(d - \frac{s}{2}\cos(\theta) - \frac{s}{2}\right)^2 + \frac{s^2}{4}\sin^2(\theta)} \qquad (5.5)$$

$$R_{24} = \sqrt{\left(d + \frac{s}{2}\cos(\theta) - \frac{s}{2}\right)^2 + \frac{s^2}{4}\sin^2(\theta)} \qquad (5.6)$$

Since it is the ratio, s/d, that is important, we define $\alpha = s/d$, factor a d from each R_{ij}, define $\hat{R}_{ij} = R_{ij}/d$, and finally write the normalized total energy as:

$$\hat{U} = \frac{4\pi d^3}{\mu_0 q^2 s^2} U = \frac{1}{\alpha^2} \sum \frac{1}{\hat{R}_{ij}}. \qquad (5.7)$$

Here, the sum is over our four particle pairs.

In Figure 5.9, we plot the normalized energy as a function of θ for two different values of α. Following Stambaugh et al. [128, 129] we choose $\alpha = 0.001$ and $\alpha = 0.75$. The small α plot corresponds to the short magnets,

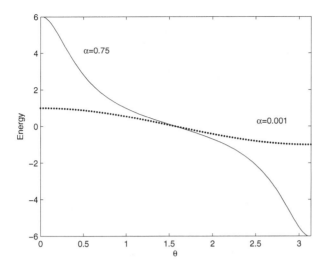

FIGURE 5.9:   The energy of a pair of touching dipoles as a function of the angle between dipole axes. Small α corresponds to short magnets, large α to long magnets. Note that $\theta = \pi$ is the preferred configuration.

while the large α corresponds to the long magnets. We see from Figure 5.9 that $\theta = \pi$ is the strongly preferred configuration for the long magnets. That is, they wish to align head to tail, forming chains. For short magnets, this preference is much weaker and side to side interactions more likely. This corresponds with the chain structures formed by the long magnets in Figure 5.7 versus the vortex structures formed by the short magnets in Figure 5.6.

The Losert group experiments and their energy based model serve as a wonderful prototypical example of an engineered self assembling system and its analysis. Structured particles, the binding force, the environment, and the driving force all play a key role. Yet, the results of the experiments, especially as a function of s/d, are not at all obvious. Their work highlights the importance of attacking the forward problem, the extreme difficulty of the backward problem, and the role of theory in the design of self-assembling systems.

5.3.2   Control Via External Fields

A natural way to extend the experiments of Sharpe [121] and Stambaugh et al. [128, 129] is to add an additional layer of control to the system by applying an external magnetic field. In the context of Table 4.1 we are altering the *environment*. The classic example of controlling a system via external magnetic fields is in the use of ferrofluids. In ferrofluids, nanoscale magnetic

FIGURE 5.10:   A ferrofluid coating a bolt suspended above a strong permanent magnet. The spikes form due to a competition between surface tension and magnetic effects. Photograph by the author.

or ferromagnetic particles are suspended in a liquid. When no external field is applied the fluid behaves as a normal Newtonian fluid. The particles are small and any inter-particle interactions are too weak to cause changes to the bulk behavior of the fluid. Upon the application of an external magnetic field however, the change in rheology is dramatic. In Figure 5.10 we show a ferrofluid in a container lying above a strong permanent magnet. At first glance, one is to be forgiven for doubting that the picture is actually of a fluid. The externally applied magnetic field has caused the suspended particles to align, form chains, and form a mesh-like structure that completely changes the behavior of the formerly Newtonian fluid. In the case of Figure 5.10 competition between surface tension and the desire of the particle chains to live along field lines accounts for the origin of the spikes.[3] Ferrofluids have found applications in shock absorbers, automobile clutches, and brakes [104].

While ferrofluids illustrate the dramatic level of control that may be obtained in a magnetic self-assembling system via the application of external magnetic fields, their is a clear gap between the intent of Sharpe [121] and Stambaugh et al. [128, 129] and studies of ferrofluidic systems. In Sharpe [121] and Stambaugh et al. [128, 129], the focus was on controlling the structures formed by the interacting particles. In ferrofluids, this structure is of secondary importance, the focus is on changes to the bulk rheology of the fluid. An experiment that bridges this gap was conducted by Golosovsky,

Saado, and Davidov [50] at the Hebrew University of Jerusalem. Golosovsky et al. embedded small disk-shaped magnets in larger styrofoam disks.[4] The styrofoam disks were then allowed to float on the surface of a liquid in a cylindrical container. A coil of wire surrounding the container was used to apply an external magnetic field to the system. The basic setup can be seen in Figure 5.11.

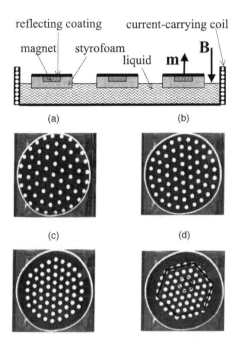

FIGURE 5.11:   The basic setup and results of the Golosovsky experiments. (a) Zero applied magnetic field. Figures (b),(c), and (d) show increasing applied field. Reprinted with permission from Golosovsky et al., Applied Physics Letters, v. 75, 1999, pp. 4168. Copyright 1999 by the American Institute of Physics.

Note that in contrast to the experiments of Sharpe [121] or Stambaugh et al. [128, 129], the magnetic dipoles of Golosovsky et al. are permanently oriented perpendicularly to the surface of the fluid. The styrofoam particles ensure that dipole orientation cannot change; the particles are unable to "flip" in the liquid. Even when no external field is applied, this causes the system to behave very differently than the Sharpe or Stambaugh systems. To understand why, imagine taking two long cylindrical magnetics and standing them on a desktop such that the north pole of each is on top. As you attempt to bring the magnets together, you'll feel a repulsive force. The magnets will want to rotate. However, if you prevent them from rotating and the magnets are free to slide in the plane of the table, as in the Golosovsky setup, this repulsive force will cause the magnets to move as far from one another as possible.

In the Golosovsky experiments 61 magnet-styrofoam particles of diameter 1.2cm were placed in a container of diameter 40cm. When no magnetic field is present, the particles self-assemble into the structure shown in Figure 5.11. In attempting to maximize their distance from one another, yet restricted by the walls of the container, the particles have formed a hexagonal lattice similar to the bubble raft system of Chapter 2. Now, when an external magnetic field pointing opposite to the particle magnetic moments is applied at the boundary of the container, the particles want to maximize both their distance from one another and their distance from the container wall. Changing the strength of the applied field changes the magnitude of the repulsive force between particles and the wall. The results can be seen in Figure 5.11. For a small magnetic field, the system moves to a state close to the zero field state, but with the whole structure slightly compressed to avoid the walls. As the field strength is increased the particles are further squeezed inwards. They maintain their hexagonal array, but as can be seen in Figure 5.11, at large applied fields, the lattice as a whole acquires a hexagonal shape.

As with the Stambaugh system, the Golosovsky experiments can be understood from an energy viewpoint. The results of Monte-Carlo simulations may be found in [50]. Here, we consider a simple model similar to that of the Stambaugh system to illustrate the role of the applied field. We consider a one-dimensional version of the experiment and treat each particle and the walls as point charges[5]. The geometry of our system is shown in Figure 5.12. We treat the walls, located at $x = \pm d/2$, as fixed and each having charge Q. By varying Q, we simulate varying the strength of the applied field. The particles are located symmetrically at $\pm s/2$ and have fixed charge q. Note that all charges are positive to correspond with the repulsive setup of Golosovsky et al. As before, we compute the Coulomb energy by considering all inter-particle pairs. We denote the particle distance by R_{ij}. From the figure we see that the total energy is

$$U = \frac{\mu_0}{4\pi} \left(\frac{qQ}{R_{12}} + \frac{qQ}{R_{13}} + \frac{q^2}{R_{23}} + \frac{qQ}{R_{24}} + \frac{qQ}{R_{34}} \right). \tag{5.8}$$

Normalizing as before, this may be considerably simplified. We find:

$$\hat{U} = \frac{\pi d}{\mu_0 q^2} U = \left(\frac{\beta}{1 - \alpha^2} + \frac{1}{4}\frac{1}{\alpha} \right). \tag{5.9}$$

Here, $\alpha = s/d$ and $\beta = Q/q$. Note that α measures the relative distance between the particles versus their distance from the wall, while β measures the relative strength of the particle charges versus the applied wall field.

In Figure 5.13 we plot the normalized energy, $\hat{U}$, as a function of α for different values of β. We see that when $\beta = 0$, the energy monotonically

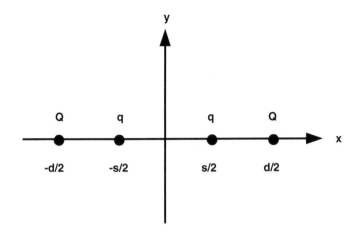

FIGURE 5.12:   The geometry of our simple one-dimensional model of the magnetic self-assembling system of Golosovsky.

decreases as a function of α. The particles want to be as far away from one another as possible. Physically, since they are constrained by the walls, they will move to a state with energy corresponding to $\alpha = 1$. When β is not zero, the energy curve changes shape and has a minimum at some value of α in $(0, 1)$. That is, the particles will maintain a distance between themselves and the walls, achieving some intermediate equilibrium.

As with the experiments of the Losert group, the Golosovsky experiments serve as a nice example of an engineered self-assembling system. In particular, these experiments illustrate how the environment may be manipulated, in this case via applied external magnetic fields, in order to alter the final state of the self-assembled system. In Chapter 7 we'll revisit the idea of manipulating the environment by applied fields when we examine self-assembling nanowires.

5.3.3   Nano-Magnets

With the advent of the technology to fabricate nanoscale particles of uniform size and type, several groups have initiated studies similar to those discussed above, but using micro and nanoscale particles. Here we briefly survey these results.

One early study by Helgesen et al. [60] used micron sized polystyrene spheres containing iron oxide and suspended in water in a study similar to that of Golosovsky et al. [50]. Their particles were uniformly sized with diameter $d = 3.6\mu$m. The water-particle suspension was placed between parallel glass plates with a gap between the plates of 5μm. An external magnetic field was applied, and the structure of particle formations observed. They found that for low applied fields, the particles tended to aggregate into clusters of short

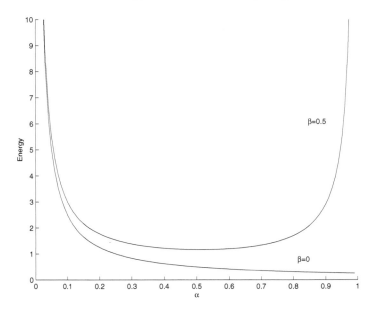

FIGURE 5.13: The normalized energy for our simple model of the Golosovsky experiments. Note that $\beta = 0$ corresponds to no applied field, while $\beta = 0.5$ models an applied magnetic field.

chains. As the field was increased, the chains became longer and the clusters separated. At the highest field levels, long chains oriented along field lines were observed. This group also performed Monte-Carlo simulations which agreed with their observations.

A more recent, but purely numerical, study was performed by Ghazali et al. [47]. This team performed Monte-Carlo simulations of nanoscale particles confined to a plane and allowed to interact via hard core repulsion and magnetic attraction. Their simulations yielded results similar to those of Helgesen et al. but also showed a dependence on particle density. For low particle densities clusters of short chains and rings were seen in the simulations. As particle density was increased, chain length increased, and at the highest densities, a vortex structure similar to that observed on the macroscale by Stambaugh et al. appeared. They also noted that if an external magnetic field were applied to the system and oriented perpendicularly to the plane of the particles, a dispersed structure, with no chains, similar to that observed by Golosovsky et al. appeared.

While work with magnetic micro and nanoparticles is in its infancy, these early results show both the utility of the macroscale studies discussed above and the promise of using magnetic particles fields as a route to self-assembly at the nanoscale.

5.3.4   Magnetic Origami

Thus far we've seen how magnetic forces can be used and manipulated to self-assemble two-dimensional structures. In an intriguing study, the group led by G.M. Whitesides[6] of Harvard University, combines the use of magnetic forces, the technology of planar fabrication, and the art of origami into an new technique for self-assembling *three-dimensional* objects [15].

Consider the sphere and its unfolding shown in Figures 5.14 and 5.15. In the most trivial sort of origami, we could cut out the planar structure of Figure 5.15, fold, and produce the three-dimensional triangulated sphere. Now, suppose that instead of us folding along the solid lines, we could place the flat structure of Figure 5.15 into some environment, introduce a binding and driving force, and induce it to self-fold into the sphere. This is the basic idea proposed in [15]. In their study, the flat shape is cut from an elastomeric sheet, magnets are embedded into the sheet to provide a binding force and the sheet is suspended in gently agitated water providing an environment and driving force to induce self-assembly.

FIGURE 5.14:   A triangulated sphere. This sphere was produced using JavaView Unfolder. This is a geometric visualization package that is freely available at www.javaview.de.

This novel biomimetic strategy provides us with examples of the types of difficult problems that arise once we attempt to self-assemble objects even slightly more complex than those considered in this chapter thus far. First, we note that we have introduced competing energies into the system. To bend our elastomeric sheet requires that we increase the elastic energy of the system. Hopefully this increase is offset by a corresponding decrease in the magnetic energy so that the system tends to the shape we desire. But, this is by no means guaranteed. Here we must solve a difficult backward

FIGURE 5.15:   The unfolding of a triangulated sphere. This unfolding was produced using JavaView Unfolder.

problem. Given our desired shape, how do we place our magnets and locate our folds so that the desired shape will be a global energy minimum of the system? At the same time we face the yield problem: How do we avoid creating local minima in which our system may get stuck? Since the projection of a three-dimensional object onto a two-dimensional sheet is not unique the answers to these questions are far from obvious. Notice that this problem is similar to the protein folding problem discussed in Part I of this text. In idealized protein folding, we have a linear chain of particles which can bond to one another. Bending the chain required an increase in elastic energy while bringing appropriate particles closer caused a decrease in binding energy. Designing the structure of the chain such that it folds into a particular shape or predicting the folded shape of a given chain presents us with precisely the same forward, backward, and yield problems that we encounter here.

The Whitesides group offered preliminary solutions to these problems in the context of their scheme via a combination of experiment, modelling, and simulation. They restricted their attention to designing two-dimensional sheets intended to fold into spherical shells. In examining a variety of cuts and patterns of magnetic dipoles they demonstrated that some systems could indeed be induced to fold correctly. However, this approach to self-assembly remains in its infancy. As with other systems we have examined and still others yet to come, the forward, backward, and yield problems for magnetic origami remain to be overcome.

5.3.5   Measuring Complexity

At the beginning of this section, we noted that the study of self-assembly leads us to questions concerning *complexity*. While in the third part of this text, we will study mathematical approaches to self-assembly and more fully address complexity questions, it is worthwhile to take a preliminary look at complexity here while the simple magnetic self-assembling systems presented above are still fresh in our minds.

The simplest and most widely used definition of complexity is due to Kolmogorov [74]. Kolmogorov considers structures that can be represented as *bit strings*, i.e., sequences of ones and zeros. To be concrete, let's consider bit strings of length ten. There are 2^{10} such possible strings and intuitively one feels that they are not all alike; some are in some sense more complex than others. For example, the string 1111111111 appears simple, while 1011001110 appears somehow more complex. Kolmogorov defines the complexity of a string as the length of the shortest computer program able to produce the string. In particular, he defines complexity as the length of the shortest computer program able to produce the string on a *Universal Turing Machine*. We'll explore Turing machines more fully later, for now, let's stick with our intuitive notions of the length of computer programs. The simple string 1111111111 could be produced by a code that in essence says:

> for $i = 1..10$
>
> WRITE 1
>
> end

On the other hand, the string 1011001110 requires more information to produce. Perhaps:

> WRITE 1
>
> WRITE 0
>
> WRITE 1
>
> WRITE 1
>
> WRITE 0
>
> WRITE 0
>
> for $i = 1..3$
>
> WRITE 1
>
> end
>
> WRITE 0

While surely this is not the absolute shortest program that could produce this output, the point is clear; the complexity of the second string, via the Kolmogorov definition, is higher.

There are two major problems with the Kolmogorov definition. First, proving that one has found the shortest program, and hence the true complexity of an object, is often difficult. The definition gives no hint of how to do so. Second, and more seriously, a purely random string has the highest complexity. Intuitively, this does not seem to measure what we are after. We would like to believe that structured objects are somehow more complex than purely random ones. In self-assembly, it is easy to build random structures, but this is not what we want, we want ordered structures.

But, for now, sticking with Kolmogorov, we see how in the study of magnetic self-assembling systems, this measure of complexity may prove useful. For example, reconsider the Sharpe experiments [121]. If we redo the experiments, this time coloring the cubes, we can see how more complex objects may be produced. Imagine that in one experiment we start with ten cubes, each with a magnet showing a north face and a magnet showing a south face. The result of the experiment will be the equivalent of the bit string 1111111111; we only have one type of "bit" or particle in our system. Now, imagine that we use two types of cubes. Five of the cubes are red and show two north faces. Five of the cubes are blue are show two south faces. If we identify the reds as ones and the blues as zeros, the result of such an experiment might be the string 1010101010, or perhaps 0101010101. We have produced a structure of higher complexity by structuring our particles differently.

The Kolmogorov idea may be used to measure the complexity of our self-assembling systems directly. Instead of looking at the output and seeking the shortest computer program that could produce that bit string, we look at the output and define the complexity as the smallest set of self-assembling particle types needed to create that structure. In the case of the magnetic cubes above, the complexity of the string 1010101010 would be 2. That is, we can build this structure using two unique particle types. This viewpoint allows us to directly measure the complexity of target structures within the context of a specific self-assembling system. We'll discuss this approach to complexity more fully in Chapter 9.

5.4   The "Cheerios Effect"

In the previous section we began with a thought experiment; take a collection of disk-shaped magnets, lay them out randomly in a container, and give the container a shake. The quote at the start of this chapter challenges us with a thought experiment of a different sort. Take a handful of Cheerios, scatter them randomly on the surface of bowl of milk, and gently tap the bowl. What happens? As observed by Jearl Walker, neighboring Cheerios rapidly pull together[7]. Given enough time all of the Cheerios cluster and form an ordered raft of floating breakfast cereal. We've formed a "cereal raft," analogous to the bubble raft of Chapter 2.

The force that drives Cheerios together in a bowl of milk, the capillary force, has been exploited by many researchers in their design of self-assembling systems. In Chapter 6, we'll explore several of these as well as several related strategies. In order to prepare ourselves to understand these efforts, in the rest of this section, we continue the analysis of the capillary force begun in Chapter 2. The important phenomenon to consider here is the difference between bubbles and floating objects. The discussion in this section draws heavily on the review papers by Kralchevsky and Nagayama [77] and by Mansfield et al. [89]. The more recent very readable paper by Vella and Mahadevan [133] is also an essential reference.

5.4.1   The Force Between Two Plates

Recall that in Chapter 2, we analyzed the system shown in Figure 5.16, and using the energy viewpoint, derived an equation for the meniscus shape, $u(x)$. We repeat this equation here for the convenience of the reader.

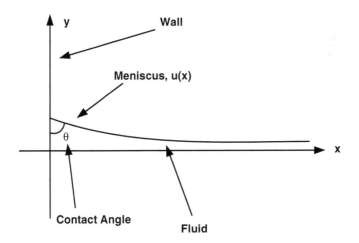

FIGURE 5.16:   The single wall meniscus system.

$$\frac{d^2u}{dx^2} - \frac{\rho g}{\sigma} = 0 \qquad (5.10)$$

Further recall that ρ denoted the fluid density, g the gravitational acceleration, and σ the surface tension. In Chapter 2, we only computed meniscus shape, and used the shape of the meniscus to give a qualitative description of why bubbles in a bubble raft felt an attractive force. Here, we extend this analysis and actually compute the force between objects in a fluid. We'll consider the idealized two plate system of Figure 5.17. We have assumed that the two

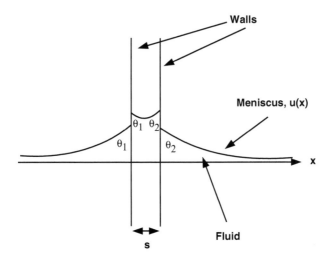

FIGURE 5.17:   The double wall meniscus system.

plates have different wettabilities, and hence the fluid-plate contact angle is different for each plate. Notice that we actually must solve Equation 5.10 three times. We must compute the shape of the meniscus to the left of the leftmost plate, in between the two plates, and to the right of the rightmost plate. We denote these three surfaces by $u_1(x)$, $u_2(x)$, and $u_3(x)$, respectively. We also assume that the plates are a distance, s, apart, and that the zero of the x coordinate is at the center of the gap between the two plates. Again, as in Chapter 2, the contact angle condition translates to

$$u_1'(-s/2) = \cot(\theta_1) \tag{5.11}$$

$$u_2'(-s/2) = -\cot(\theta_1) \tag{5.12}$$

$$u_2'(s/2) = \cot(\theta_2) \tag{5.13}$$

$$u_3'(s/2) = -\cot(\theta_2) \tag{5.14}$$

We'll also require that the disturbance to the interface vanishes as $x \longrightarrow \pm\infty$. Solving Equation 5.10 and applying the boundary conditions in each region, we find

$$u_1(x) = L \cot(\theta_1) \exp(\frac{1}{L}(x + s/2)) \tag{5.15}$$

$$u_3(x) = L \cot(\theta_2) \exp(-\frac{1}{L}(x - s/2)) \tag{5.16}$$

$$u_2(x) = \frac{L}{\sinh(s/L)}(\cot(\theta_1) \cosh(\frac{1}{L}(x - s/2)) + \tag{5.17}$$

$$\cot(\theta_2) \cosh(\frac{1}{L}(x + s/2))).$$

Here we have once again used L to denote the capillary length, i.e.,

$$L = \sqrt{\frac{\sigma}{\rho g}}.$$ (5.18)

Now that we have solved for the shape of the meniscus in each region, we can compute the force between the plates. There are two ways to proceed. In [133], Vella and Mahadevan note that this force can be computed by finding the difference in pressure along both sides of one of the plates. Alternatively, we could compute the total energy of the system and differentiate with respect to s in order to compute the force. In either case, we find that the force between the plates can be reduced to

$$\frac{\rho g}{2}(u_1(-s/2)^2 - u_2(-s/2)^2).$$ (5.19)

This can be evaluated to yield

$$F = \frac{\sigma}{2}\left(\cot^2(\theta_1) - \frac{(\cot(\theta_1)\cosh(s/L) + \cot(\theta_2))^2}{\sinh^2(s/L)}\right).$$ (5.20)

Here, we have assumed a negative force corresponds to attraction and a positive force corresponds to repulsion. Note that this expression depends on both θ_1 and θ_2. Further, the sign of this expression varies according as the values of θ_1 and θ_2. The sign of Equation 5.20 also depends upon the dimensionless ratio s/L, that is the relative values of the plate gap versus the capillary length. Several situations may occur. Typical is the situation shown in Figure 5.18. Here we have plotted the force for fixed values of the contact angles as a function of the ratio s/L. We see that the force is positive when s/L is large, the plates push away from one another, but that the force is negative, or attractive, for small values of s/L. This ability of the system to exhibit both attractive and repulsive forces, as well as a combination of the two, is the main point of this analysis.

5.4.2   Floating Particles and Other Forces

Thus far we've seen, in both our bubble raft system and in the analysis above, that the surface tension of a liquid can provide attractive and repulsive forces between objects in that liquid. However, we still have not dealt with genuinely floating particles. The key difference between the analysis above and the case of floating particles lies in the fact that vertical as well as horizontal forces must be considered. If an object, such as a paper clip, is floating on the surface of a liquid, forces in the vertical direction must be sufficient to keep it there. Or, in terms of energy, the energy gained by reducing the gravitational potential of the paper clip, is not offset by the increase in surface energy expended as the liquid stretches to accommodate the sinking clip. Hence, the paper clip floats.

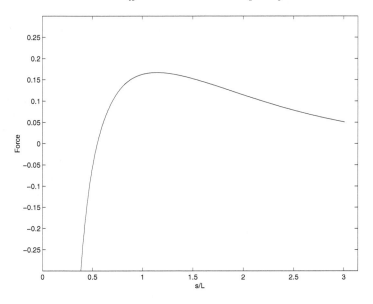

FIGURE 5.18: The force between our plates as a function of the dimensionless gap, s/L.

But the fact that the clip floats implies that some component of the surface forces are acting in the vertical direction and this must be accounted for in any analysis of a floating particle. In principle, such an analysis is not any more difficult than the analysis of the previous section. In practice, the geometry often forces us to resort to numerical computations. In certain idealized geometries, the force between floating objects can be obtained via an asymptotic analysis. For spherical particles of radii R_1 and R_2 placed a distance L apart, Kralchevsky et al. [77] carried out such an analysis. They found the force could be written as

$$F = -2\pi\sigma Q_1 Q_2 q K_1(qL)[1 + O(q^2 R_k^2)]. \tag{5.21}$$

Here, σ is the surface tension, $Q_i = r_i \sin(\psi_i)$, r_i is the radius of the ith particle's contact line, ψ_i is the contact angle for the ith particle, and

$$q^2 = \frac{\triangle pg}{\sigma}. \tag{5.22}$$

The term $\triangle p$ is the difference in the mass densities between the fluid on which the particle is floating and the fluid above. The Q_i are often referred to as *wettability coefficients*. Finally, K_1 is the modified Bessel function of order one.

On this basis of this computation for the force between particles at the surface of a fluid, we can identify several situations of practical importance.

These are shown in Figure 5.19. Note it is the combination of the wetting properties of the spheres and their relative densities that determine whether forces will be attractive or repulsive. In Figure 5.19 (a) we have an attractive

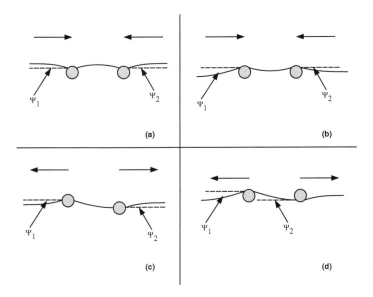

FIGURE 5.19:   The force between floating spheres for different configurations of density and wetting properties.

force between two floating particles. Here, the condition $\sin(\psi_1)\sin(\psi_2) > 0$ must be satisfied. When this product becomes negative the force becomes repulsive as in Figure 5.19 (c). In Figure 5.19 (b) and (d), we have immersed particles exhibiting both attraction and repulsion. Again the attractive case requires $\sin(\psi_1)\sin(\psi_2) > 0$ while the repulsive case requires $\sin(\psi_1)\sin(\psi_2) < 0$.

The reader who is uncomfortable with this sort of analysis might find it useful to think solely in terms of surface energy when considering a surface tension driven self-assembling system. Simply bear in mind that the system must expend energy in order to increase the area of its surface. If this area can be reduced by bringing particles together or pushing them apart, the system will generally do so. If this type of thinking leads you astray in your analysis of a given system, the problem probably lies in the contribution of those messy vertical forces or in the details of the meniscus shape and you'll have to immerse yourself in calculations like those given above.

Try It Yourself - Self-Assembling Soda Straws

In this chapter we've seen how simple forces, exploited cleverly, can lead to relatively complex self-assembled structures. Capillary forces are one of the primary examples of such forces and in addition to being useful in engineering, nicely lend themselves to easy, "kitchen-sink" style experiments. One such experiment, first proposed by Campbell et. al. in [25], uses ordinary soda straws to demonstrate capillary driven self-assembly. To carry out this experiment yourself you will need:

Materials

- Beaker, Petri-dish, or other large-mouthed container

- Soda straws (One type is sufficient, but experimenting with multiple types is interesting.)

- Scissors

- Orbital shaker (This is optional, shaking by hand works fine.)

Procedure Begin by cutting the soda straws into short pieces of equal length. The number of pieces you need depends on the size of your container. Start with about ten and add from there. One and one-half centimeter long pieces work well. The cut should be perpendicular to the axis of the straw. Fill your beaker or Petri dish with enough water to submerge the soda straws. Push each tube under the water, forcing the air out, and allow it to return to the surface. Gently agitate the container to force the system out of a local equilibrium. The straws should assemble into a pattern similar to that shown in Figure 5.4.2.

Things to Try As discussed in this chapter the curvature of the meniscus between the floating soda straws determines whether there will be an attractive force or repulsive force between them. Even in a system as simple as water and soda straws, this force, and the resulting interactions can be manipulated. Some ideas for further experiments are:

- Vary the energy with which you shake the vessel. How does this effect the approach to equilibrium?

- Use a variety of soda straws with different diameters. How do they interact?

- Add soap to the water to decrease the surface tension. What happens to assembled rafts?

- Cut the soda straws in half, or cut the ends at an angle. How does this change the interactions? The final shape?

- Build soda straw "bugs" by connecting pieces at an angle prior to placing them in your vessel. Can you create more complex self-assembling structures?

Further Reading The original article by Campbell et al. [25] contains a nice discussion of the energetics of the soda straw system. The supplemental material for the article available on the *Journal of Chemical Education* web page has other suggestions for extensions of the experiment. A similar system, using "Kix" breakfast cereal and suggested as an alternative to the bubble raft is discussed in [36]. The University of Wisconsin's Materials Research Science and Engineering Center on Nanostructured Interfaces' web page (mrsec.wisc.edu) features several nice movies of systems self-assembling under the action of capillary forces.

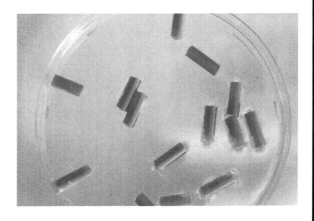

FIGURE 5.20:
Soda straws aggregating into regular "raft-like" structures on water in a Petri dish. This is the starting disordered state. Photograph by the author.

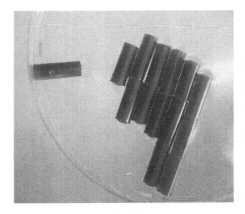

FIGURE 5.21: Soda straws after aggregation has occurred. Photograph by the author.

Profile - Saul T. Griffith

All of the individuals profiled in this book may be rightfully called "tinkerers" and "inventors." But, none of the individuals profiled here fits that description more accurately than Saul T. Griffith.

Griffith, a product of MIT's famous Media Labs, holds degrees in engineering as well as a doctorate in media arts and sciences. In his thesis, "Growing Machines," Griffith explored the question of how self-assembled systems might be programmed to build complex structures. As part of this work, Griffith designed and built a set of magnetic tiles that when arranged in the proper sequence could be induced to fold into any two dimensional shape composed of square pixels. The output of one such experiment, a sequence that folds into MIT's logo, is shown in Figure 5.22.

Griffith's tinkering extends beyond self-assembly. He is the co-founder of Low Cost Eyeglasses, a company committed to bringing affordable prescription eyecare to rural and developing communities. The foundation of this company is Griffith's novel device that allows any eyeglass lens to be produced from a single flexible pair of molds. For this work, Griffith received the National Inventor's Hall of Fame Collegiate Inventors Award.

But still, eyecare and self-assembly are just the tip of Griffith's tinkering iceberg. Griffith, barely past thirty, already holds a range of patents spanning optics, nanotechnology, and materials science. For fun, Griffith engages in the somewhat arcane sport of "kitesurfing," of course, building his own large inflatable wings and dragging himself across ice, water, or whatever large open surface he can find. In addition, Griffith has a deep commitment to developing new ways to communicate science to children. Together with an MIT colleague, Joost Bonsen, Griffith founded "Howtoons," a novel comic strip series that presents science to children in a hands-on and entertaining way. He's even sketched one on self-assembly; you can read it in Color Plate 11.1.

FIGURE 5.22: Saul Griffith's self-folding magnetic tiles folded into the MIT logo. Credit: Saul T. Griffith.

5.5   Chapter Highlights

- Nature utilizes four key components in creating self-assembling systems. These are: structured particles, binding forces, the proper environment, and a driving force. Man-made systems utilize the same four components.

- The *forward problem* of self-assembly refers to predicting the structures formed given a set of structured particles, a binding force, an environment, and a driving force.

- The *backward problem* of self-assembly refers to determining what set of particles, binding forces, environments, and driving forces will produce a desired outcome.

- The *yield problem* of self assembly refers to determining what fraction of the components in a self-assembling system actually form the desired structure.

- Penrose and Penrose [100] showed that a very simple system made from common materials can exhibit self-assembling properties.

- Experiments with magnetic systems provide an easy framework for understanding the forward, backward, and yield problems. Magnetic systems also have practical applications.

- The "Cheerios Effect" is the tendency for small particles, such as breakfast cereal, to clump together when floating on the surface of a liquid.

- Capillary forces, such as those responsible for the Cheerios Effect, may be tailored by changing the wetting properties of the system.

5.6   Exercises

Section 5.2

1. The Penrose Model may be simulated on a computer. Assume you are given an initial random string of components of length n. Assume an A-B complex is added to the left end of the string. Now, assume that in each time step the first possible A-B complex to the right of the seed is formed. Under these assumptions write a computer program to simulate the action of the Penrose Model. For a typical random string, how many time steps are necessary to completely assemble all possible new complexes?

2. Can you design a more complex version of the Penrose model? Can you do so while preserving the simplicity of the construction and design?

3. Can you design a 2-d version of the Penrose Model? Perhaps the particles could be placed on an air-hockey table and allowed to interact randomly? Eric Klavins has exploited the air-hockey idea to great effect. See the profile in Chapter 9.

Section 5.3

4. The self-assembly of magnetic dipoles may be simulated simply. Imagine a large box of individual magnets. At random, select two magnets from the box, join them together, and replace them in the box. Repeat. If this process is simulated on a computer, how does the result compare to what is actually seen in a real experiment? Why?

5. The Sharpe experiments may be extended in several simple ways. Imagine that the cubes are colored. Say, dipoles are blue, monopoles red, and end caps green. How does this add to the complexity of the self-assembled structures?

6. Extend the energy arguments used to analyze the experiments of Stambaugh et al. to the case of three contacting particles. This time, instead of an energy curve, you will have an energy surface. What will the system do? How does it vary with s/d?

Section 5.4

7. It is sometimes hard to believe that something as fragile as the surface of a liquid can exert a significant force. To demonstrate this, take a stiff wire and a piece of string and construct the U-shaped system of Figure 5.23. Dip this into a tray of soap solution. What happens? What shape does the string assume?

8. Consider the experiment of the previous problem. Write down the energy of this system and use this to mathematically determine the shape assumed by the string. This problem requires knowledge of the calculus of variations. See the book by Oprea in the Related Reading section at the end of Chapter 2 for a full discussion.

9. Many demonstrations of the effects of surface tension have been devised. One classic involves pepper floating of the surface of water. Fill two Petri dishes with water and sprinkle pepper on their surface. In one place a single drop of rubbing alcohol at the center. In the other place a single drop of liquid soap. What happens to the pepper in each case? Why?

10. Another demonstration of the effects of surface tension involves toothpicks and a drop of water. If we start with an array of toothpicks

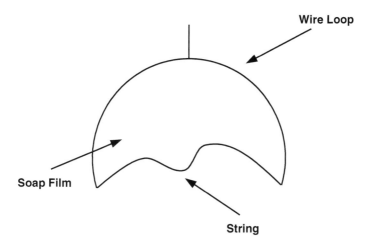

FIGURE 5.23: The wire loop, string, and soap film setup for Exercises 7 and 8.

arranged as in Figure 5.24 (a), a drop of water placed at the center of the arrangement will cause the toothpicks to move to the configuration of Figure 5.24 (b). Explain why this happens. You might want to try it for yourself.

11. Repeat the calculation of the force between two plates immersed in a fluid for the case when the plates are at an angle to the flat interface. How does this angle control whether the plates are attracting or repelling? How might this be used in designing an artificial amphiphilic particle?

5.7   Related Reading

The quote at the start of this chapter is from the entertaining book by Jearl Walker. One could spend a lifetime contemplating the questions he presents.

J. Walker, *The Flying Circus of Physics*, John Wiley and Sons, 1975.

A variety of ingenious self-replicating mechanical and electromechanical devices are described in:

R.A. Freitas Jr. and R.C. Merkle, *Kinematic Self-Replicating Machines*, Landes Bioscience, 2004.

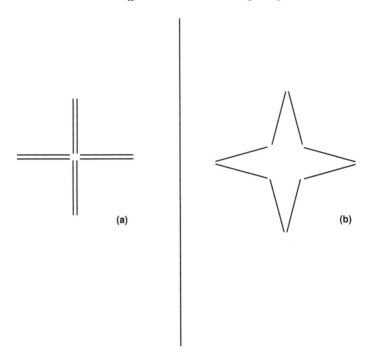

FIGURE 5.24: The toothpick array for Exercise 10.

5.8 Notes

1. A more common terminology would be to call this the *inverse problem* of self-assembly. I've chosen to use the term backward rather than inverse throughout this book to attempt to avoid confusion with the numerous connotations of the term inverse in science, engineering, and mathematics. Of course, the term backward has its own set of connotations.

2. These videos may be viewed on the website for this text.

3. Ferrofluids are relatively easy to make and fun to play with, although quite messy. Sources of ferrofluids may be found on the web page for this book.

4. Thereby reviving the system criticized by Bragg and Nye in their original bubble raft paper. See Chapter 2.

5. This one-dimensional problem when extended to N particles yields a surprising result. The particles distribute themselves on the line at precisely the zeros of the Nth Chebyshev polynomial. This problem, and the equivalent version on a spherical surface has connections to cryptography.

6. See profile in Chapter 6.

7. For an example of the Cheerios Effect taken to the extreme, see the cartoon by Saul Griffith in the color plates.

Chapter 6

Static Self-Assembly

Fundamentally, all sciences reason in the same way and aim at the same object. They all try to reach knowledge of the law of phenomena, so as to foresee, vary, or master phenomena.

Claude Bernard, *Experimental Medicine*

6.1 Introduction

From the engineering perspective, the best understood and most highly developed engineered self-assembling systems are the *static* self-assembling systems. Static self-assembling systems are those that operate via a principle of energy minimization. A collection of particles slides down an energy gradient reaching a global or local minimum, ending in some final ordered configuration. The system then remains at this thermodynamic equilibrium and the end product is clear for all to see.

The closest counterparts in the natural world of engineered static self-assembling systems are crystals. In fact, many of the systems examined in this chapter are often discussed and sometimes dismissed as examples of artificial crystallization. While the criticism of these systems as examples of "mere crystallization" does have some merit, and while our ultimate goals in designing future systems are loftier than those achieved here, it is amazing to see just how much can be and has already been accomplished via a study of artificial crystallization.

The relative wealth of examples of static self-assembling systems in the literature means that in this chapter we must make difficult choices. There is simply not room to cover the entire spectrum of what has been accomplished. So, here, we will focus on four broad themes: systems that bond using capillary forces, template driven systems, systems that create structures by minimizing surface energy, and systems that assemble by folding.

We begin in Section 6.2 with a discussion of systems that use capillary forces to create particle-particle bonds. Of course, we've already encountered systems like these before, specifically the bubble raft of Chapter 2 and the Cheerios system of Chapter 5. The systems of this section operate in much the same way that the bubble raft and the Cheerios system operate and we'll find that the analysis of those systems allows us to understand the systems presented here quite easily. However, the examples of Section 6.2 go far beyond breakfast cereal. We'll see how adding structure to floating particles, the "Cheerios," allows for the formation of patterns much richer than a hexagonal bubble raft. We'll also see how the idea of assembling via capillary forces can be extended to allow the creation of three dimensional functional devices. Finally, at the end of Section 6.2, we'll explore two systems that push the use of capillary forces to their limits. The first allows for reconfiguration of assembled structures in response to changes in the environment. The second explores how capillary binding can be used to produce systems that compute.

In Section 6.3 we examine systems that use templates. We focus on two different systems. In this first, the use of the capillary bond is coupled with the use of a template to produce finite arrays of particles with a predefined shape. Here, the use of a template allows one to deal with the backward problem of self-assembly. The configuration of the final structure is specified by the shape of the template. The challenge is then reduced to designing particles that will fit into this template. In the second, a template is used as an intermediate step in a self-assembly process. Here, the template is used first to create, and then to align artificial amphiphiles. The amphiphiles self-assemble in a process mimicking that of micelles, but here, the alignment by the template turns out to be necessary in order for self-assembly to occur.

In Section 6.4 we examine efforts to create structures such as microchannels, waveguides, and microlenses, using nature's tendency to minimize surface area. This brings the subject of self-assembly into contact with the mathematical theories of minimal surfaces and constant mean curvature surfaces. We'll explore one of the simplest examples of a minimal surface, the catenoid, and show how its shape may be computed from an energy minimization principle. We then discuss experimental efforts in this area, discuss how they connect to the mathematical theory, and finally, we'll encounter the phenomenon of symmetry breaking.

In the final section of this chapter, Section 6.5, we examine self-assembled systems that take inspiration from nature's folding of proteins. We saw in Chapter 3, that nature utilized folding to create complex functional structures such as the ribosome. In Section 6.5 we'll examine a simple system that uses folding and that has been both analyzed theoretically and realized experimentally. We'll see how constraining the system and introducing the idea of linear sequential folding allows one to effectively solve the forward, backward, and yield problems of self-assembly.

6.2 Assembly via Capillary Forces

The earliest attempt to use capillary forces to create an engineered self-assembling system was carried out by Hosokawa et al. in 1996 [62]. Using standard MEMS fabrication techniques the group constructed a set of particles with the shape shown in Figure 6.1. Their work illustrates how structured particles may be designed to produce a desired target structure. In their work they encountered the forward, backward, and yield problems described in Chapter 5.

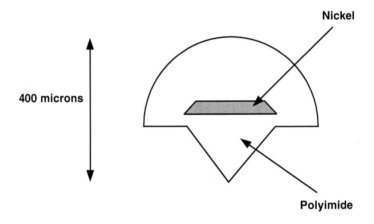

FIGURE 6.1: The basic particle in the Hosokawa system.

The group fabricated 108 polyimide particles about 400 microns in diameter. Each particle contained a strip of nickel as indicated in Figure 6.1. The particles were allowed to float on the surface of water and were driven by the application of an external magnetic field. Polyimide is non-wetting and the group noted the presence of a highly curved meniscus surrounding their particles. They further noted that near the sharp tip of their particles, the meniscus gradient was very large. Since, as we've seen, the surface energy of the surrounding water is proportional to the surface area, this implies that most of the energy expended by the fluid in creating new surface is expended near the sharp tips. In fact, the group chose to design their particles to exploit this behavior. If two particles with the shape shown in Figure 6.1 are nearby, capillary forces will draw them together. However, energetically it is most favorable for the particles to align with their tips in contact; this orientation reduces the surface energy as much as is possible. Based on this, the group set the experimental goal of assembling structures like those shown in Figure 6.2. Notice that Hosokawa et al. were grappling with the forward and

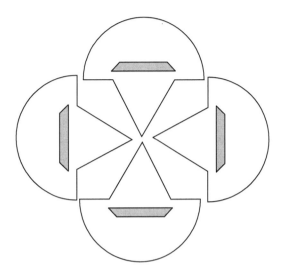

FIGURE 6.2:   The target structure in the Hosokawa system.

backward problems of self-assembly. In constructing particles with the shape of Figure 6.1 and observing that interfacial energy was maximized near the tip, they were solving the forward problem. That is, they were attempting to predict how the particles would assemble. In setting the goal of Figure 6.2, Hosokawa et al. were attacking the backward problem. Having identified their target structure they attempted to design their particles such that they would assemble into the given structure.

Unfortunately, as is typical in self-assembly, local minima got in the way. In their first set of experiments, the group discovered what they termed "reverse coupling." This is shown in Figure 6.3. This structure arises because there is also large interfacial energy expended by the corners of the semi-circular part of their particles. The reduction in energy achieved by the system in Figure 6.3 represents a local energy minimum. Particles became trapped in this local energy minimum and were unable to find their way to the global minimum of Figure 6.2. Hosokawa et al. had encountered the yield problem. Their experimental system, consisting of 108 particles, yielded only eight complete target structures in the most successful of their attempts. This represents a yield of about 30 percent.

In order to overcome these difficulties Hosokawa et al. redesigned their particles. This time they combined both polyimide and polysilicon layers to again

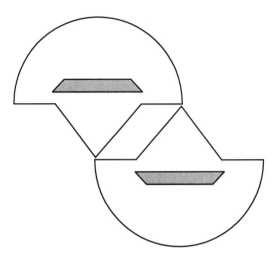

FIGURE 6.3:   A typical binding defect in the Hosokawa system.

form structures like those shown in Figure 6.1. However, this time, the particles were not planar. Rather, a curl was induced in the sharp tips. When these particles were floated on the surface of water, the height of the tips was different than the height of the corners of the semi-circular part of the structure. This design allowed them to make the attractive capillary forces between the tips and semi-circular corners become repulsive. This meant that the reverse coupling structure of Figure 6.3 could be eliminated as a local energy minimum. Unfortunately, this still did not increase the yield of the final target structure. This time, the group found that when three of the particles had joined a target structure, the fourth particle rarely joined, presumably because the repulsive forces were now overcoming the tip to tip attractive force.

This early study by Hosokawa et al. nicely illustrates how the capillary bond may be used in the design of a self-assembling system. Their work also highlights how the forward, backward, and yield problems arise in capillary driven self-assembly experiments. In addition, the group also formulated a mathematical model of their experimental system. Their model, based on reaction kinetics, is discussed further in Chapter 9.

6.2.1   Assembly at a Liquid-Liquid Interface

A sequence of studies, similar to the work by Hosokawa et al. , was carried out by the group led by G.M. Whitesides [17, 18]. The Whitesides group introduced two key innovations that had not been explored by Hosokawa et al. First, they replaced the water-air interface of the Hosokawa study with a liquid-liquid interface. Their interface was formed by placing a layer of water on top of a layer of perfluorodecalin ($C_{10}F_{18}$). The density of these two fluid layers could be adjusted relative to the density of their particles. This gave them an additional measure of control in their system; by adjusting this density, they could adjust where the particles sat relative to the flat fluid-fluid interface. Second, their particles were fabricated from polydimethylsiloxane (PDMS). The use of PDMS allowed them to achieve fine control over the wetting properties of the particles. The surfaces of PDMS are naturally hydrophobic, but can be made hydrophilic by oxidation. Selective oxidation of the particles allowed the Whitesides group to create a wetting pattern on the surface of the particles in turn allowing them control over the final assembled structures.

Their initial particles were hexagonally shaped, between 1mm and 10mm in length, and between 0.5mm and 3mm thick. The top surfaces of the particles were made hydrophilic, while the bottoms were made hydrophobic. This ensured that the particles would remain at the water-$C_{10}F_{18}$ interface. The system was placed on an orbital shaker and driven for several hours. In their first set of experiments, the six sides of the hexagonally shaped particles were all allowed to remain hydrophobic. This meant that when floating at the interface, the particles were wetted by the $C_{10}F_{18}$, producing curved menisci. In turn, this produced an attractive capillary force between the particles. The result of this experiment is shown in Figure 6.4 (c). Note that hexagons tile the plane, and the structure assembled reflects this fact. A large, simple, crystalline structure had been produced. In the next set of experiments, the sides of the hexagon were alternately hydrophillic and hydrophobic. This meant that there was an attractive force between the hydrophobic sides, but none between the hydrophillic sides. The result of this experiment is shown in Figure 6.4 (b). Here, the hexagons have formed an open structure, bonding with their neighbors on three faces, but remaining in contact with liquid on the other three. In another variation of this experiment, the group constructed cross-shaped particles as shown in Figure 6.4 (a). Here, the flat edges of the cross were made hydrophobic while the curved edges were allowed to remain hydrophillic. Once again, an open crystalline structure was formed, this time with circular vacancies.

These experiments again illustrate the backward, forward, and yield problems in self-assembly. Through fine control of the wettability of their particles the group was able to solve the backward and forward problems for their systems. The yield in these experiments was also essentially one-hundred percent. However, there is an important difference in the goals of the Hosokawa group and the Whitesides group. Recall that the Hosokawa group attempted to create a final product that was *finite* in extent. This meant that once their target structure was formed it had to stop self-assembling. In the Whitesides experiments, the final product was *infinite* in extent. It only stopped growing when the system ran out of particles or space. The difference between these two experiments is one of *information*.

Profile - George M. Whitesides

In reading this chapter you've undoubtedly noticed that many of the most interesting self-assembly experiments to be performed in recent years were carried out by the group led by George M. Whitesides. The quantity and quality of the publications from the Whitesides group should come as no surprise. Whitesides holds the distinction of being one of the most prolific *and* one of the most highly cited chemists in the world.

Presently a Woodford L. and Ann A. Flowers University Professor at Harvard University, Whitesides was trained as a chemist. After receiving an A.B. degree from Harvard in 1960, Whitesides migrated west and received a Ph.D. from the California Institute of Technology in 1964. Returning to the east coast, Whitesides served on the faculty of the Massachusetts Institute of Technology for almost twenty years. Finally, in 1982, Whitesides returned to his alma mater and joined the faculty at Harvard University.

In a 2002 interview with Thomson Scientific's *Science Watch* newsletter, Whitesides was asked if he has a personal "Holy Grail" for his research. He answered:

> *Yes. To be able to make complex systems, either structurally or functionally, by self-assembly. What we have at the moment is a fairly simple set of structures. We would like to develop a synthesis technology that would enable the making of nanometer-scale, three-dimensional structures on surfaces with arbitrarily chosen properties. It's materials by design.*

The Whitesides group has already brought us closer to this goal and surely will continue to do so in the years to come.

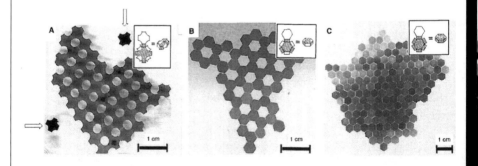

FIGURE 6.4: Mesoscale artificial crystals generated by the Whitesides group. From Bowden, et al., Science, v. 276, pp. 233-235, (1997), Reprinted with permission from the AAAS.

More information had to be encoded in the Hosokawa particles than in the Whitesides particles. The yield effectiveness of these two systems can be directly traced to this difference in information.

The Whitesides and Hosokawa experiments also illustrate the role of the driving force in self-assembly. In the Hosokawa experiments the driving force was achieved through the application of an external magnetic field. In the Whitesides experiments the driving force was accomplished by mechanical shaking. In either case, the application of a driving force created random motion of the particles and hence random opportunities for the particles to interact. Note that the strength of the agitation affected the capillary bond. As shaking frequency was increased the average interaction between particles switched from bonding to disassociation. Hence, shaking in a capillary bond system is the analog of temperature in a chemical bond system.

The Whitesides group also conducted extensive experiments with other particle structures. A lock and key design allowed them to assemble linear chains. An example of this is shown in Figure 6.5. The group also constructed particles of different heights. This introduced an additional method of control over the strength of the capillary interactions. Finally, the group compared the use of lightweight particles interacting via their hydrophobic faces with heavy particles interacting via their hydrophillic faces. By adjusting the density of their fluid layers the group could control whether the particles were "heavy" or "light." Details of these experiments may be found in [17, 18].

6.2.2 Capillary Forces and Three Dimensional Structures

The studies discussed above concerned the creation of two dimensional self-assembled structures. In another series of experiments, the Whitesides group extended the use of the capillary bond to create structures in *three dimensions* [53, 22, 16, 63, 139, 23]. In addition, their three dimensional structures were often *functional*. The group embedded electrical circuits in each of their particles and allowed the capillary bond to both join particles and make electrical connections.

The basic idea behind many of their three dimensional experiments is shown in Figure 6.6. The group fabricated truncated octahedral particles from a photocurable polymer. Circuit components could be embedded within these particles. The square faces of the octahedra were used to make particle-particle and electrical connections. A patterned copper-polyimide sheet was glued to these square faces. The octahedra were then dipped in liquid solder. The solder adhered to regions of exposed copper. A collection of these octahedra were immersed in a heated solution of potassium bromide. The solution was approximately isodense, so that the particles were effectively neutrally buoyant. The entire mixture was then gently agitated and the particles allowed to self-assemble. In this case, assembly was driven by capillary interactions between the liquid solder droplets. As the solder attempted to minimize its free surface area, the particles were drawn together and a capillary bond created.

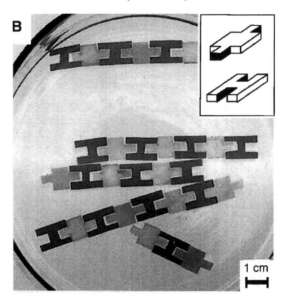

FIGURE 6.5:   Lock and key self-assembly from the Whitesides group. Dark faces indicate hydrophobic sides, light face indicate hydrophillic sides. From Bowden, et al., Science, v. 276, pp. 233-235, (1997), Reprinted with permission from the AAAS.

Fully assembled functioning structures and the pattern of solder interconnects may be seen in Color Plate 11.10.

From a design point of view, the pattern of solder dots on the square faces represents the most interesting and challenging aspect of this work. To understand this, let's first imagine that we have two square faces, each with a single circular drop of solder placed at their center as in Figure 6.7 (a). We can imagine sketching a one-dimensional energy landscape for this system by restraining the motion of the squares to an axis running through the center of the dots, i. e., the dashed line in Figure 6.7. As the two squares move together, the energy will remain flat. Once the solder dots begin to overlap the energy decreases, reaching a global minimum when the dots are perfectly aligned. Note that this is a one-dimensional picture, in the full energy landscape there is a rotational freedom, the squares need not be aligned with their edges matching. This symmetry could be broken by using square dots. But nonetheless, the centers are driven to align by capillary forces. Now, imagine a pair of squares with two solder dots, placed like those in Figure 6.7 (b). This time, the global energy minimum, where the solder dots sit on top of one another, does cause the edges to align. However, we have also created two local energy minima. As the squares move towards one another, the energy remains constant until the first pair of dots begins to overlap. The energy

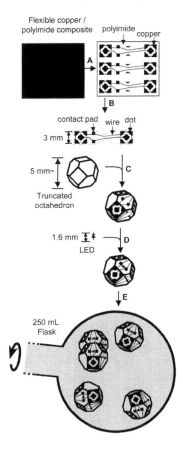

FIGURE 6.6:   The Whitesides system for forming three dimensional electrical networks by self-assembly. From Gracias, et al., Science, v. 289, pp. 1170-1172, (2000), Reprinted with permission from the AAAS.

then decreases to a local minimum. Pushing the squares further requires that we *increase* the energy of the system. If we did manage to do so, we could reach the global minimum where both dots were aligned, but in practice, the system will tend to remain at the local minimum. To overcome this creation of local minima the Whitesides group had to carefully design the pattern of solder on their copper faces. Their design, arrived at by heuristic reasoning, is shown in Figure 6.8. If we imagine two faces like the ones in Figure 6.8 coming into contact, we see that however we orient the faces, the system can reduce the interfacial energy by moving the squares still further together. That is, we've created an energy gradient and the particles slide down this gradient towards the desired global minimum. The difficulties of Figure 6.7 presented by the presence of local minima have been overcome.

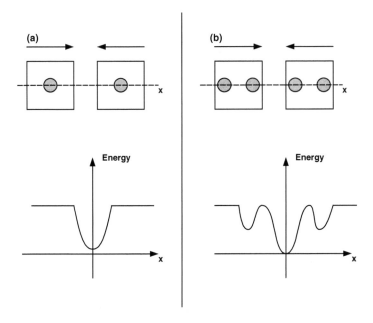

FIGURE 6.7: One dimensional energy landscapes for a solder dot system. In (a) we have a single solder dot and a single minimum in the energy landscape. In (b), the presence of two dots creates two local minima in addition to a global minimum.

While extremely clever, we note again that the pattern of Figure 6.8 was designed heuristically. This presents a well-posed fascinating challenge to the self-assembly researcher that shares many features with the more complicated protein folding problem. The question can be posed: How do we design a pattern of n solder dots on our square face so that the system has a global minimum when two faces are correctly aligned and so that we avoid the creation of local minima? Further practical constraints can be placed on the problem. For example we can require that no two dots are closer than a given distance, that the amount of solder used is minimized, or that a dot must have a given minimum area. Whatever variations of the problem we pose, the essence remains the same – this is a wonderful example of an easy to grasp, but difficult to solve, instance of the backward problem in self-assembly.

Efforts by the Whitesides group to self-assemble three dimensional structures goes beyond what has been described here. We refer the reader to [53, 22, 16, 63, 139, 23] for further details.

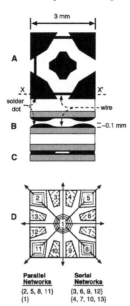

FIGURE 6.8:   The pattern of solder dots used by the Whitesides group to reduce the presence of local minima in the energy landscape. From Gracias, et al., Science, v. 289, pp. 1170-1172, (2000), Reprinted with permission from the AAAS.

6.2.3   Reconfigurable Capillary Driven Systems

A classic geometric dissection problem asks: Given the dissection of the triangle in Figure 6.9, can the pieces be rearranged to form a square? In 2002, the Whitesides group asked the related question: Given the dissection of the triangle in Figure 6.9, can the pieces be designed so that they will rearrange *themselves* to form a square?

As in their earlier studies, the group began by fabricating PDMS tiles designed to float at a water-perfluorodecalin interface. Again, as before, the edges of their tiles were patterned to be either hydrophillic or hydrophobic. This time however, they allowed themselves to adjust the density of the water phase in their two phase system. By controlling this density they could force the particles to float with their center of gravity below or above the water-perfluorodecalin interface. If the particles floated below the interface, capillary forces between the hydrophillic faces were significant while those between the hydrophobic faces were negligible. If the particles floated above, the situation was reversed. In this manner, the group encoded two different target structures into the same set of particles. Adjustments to the density allowed them to select which shape would be assembled. By changing the

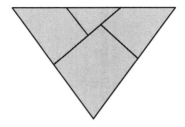

FIGURE 6.9: A classic dissection puzzle. Can the pieces be rearranged to form a square?

density they could force a set of particles in the configuration of Figure 6.9 to flip into a square configuration. In this study, the group also encountered a new problem that required them to introduce the idea of tethered particles. If particles like those in Figure 6.9 are allowed to freely assemble they generally choose a local minima consisting of small two particle aggregates. To ensure that their system reliably chose the square or the triangle, the particles were tied together in a chain using small cotton threads. This effectively turned a random assembly process into a linear folding process. Details of the design of this system may be found in [91].

6.2.4 Computing with Capillary Forces

In 2000, Paul W.K. Rothemund wondered if the capillary driven systems pioneered by Hosokawa et al. and explored by Whitesides and others could be driven to do more than simply form large crystals. In particular, he wondered if such systems could be made to *compute* [106, 108]. Rothemund realized that structures such as those in Figure 6.4 are actually examples of *tilings* or partial tilings of the plane. He further knew that through the concept of a Universal Turing Machine, tiling had been shown to be equivalent to computation. So, why couldn't a capillary driven system be made to compute?

To begin to understand how a purely mechanical device can compute, let's focus on simple binary operations. If you are even vaguely familiar with computation or with elementary logic, you are probably familiar with simple operations such as AND, OR, and NOT. Rothemund focused on the binary operation known as XOR, the so-called exclusive OR. This operation takes two arguments as its input, compares them, and if either is true, but not both of them, returns the answer TRUE. Otherwise, the XOR operator returns the answer FALSE. Using ones to denote TRUE and zeros to denote FALSE, we may summarize the operation of XOR in a table. This summary appears in Table 6.1.

Now, to encode the XOR operator in a capillary bond driven system, Rothemund designed and fabricated four different types of tiles. The lower half of

TABLE 6.1: The
Logical Operation XOR

Input	Output
0 0	0
0 1	1
1 0	1
1 1	0

each tile was designed to bind to an appropriate input string, while the upper half was designed to represent the output of the XOR operation. We've shown a simplified abstraction of Rothemund's four tile types in Figure 6.10. In practice, Rothemund found that he had to use a combination of complex

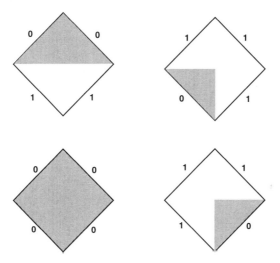

FIGURE 6.10: A simplified version of Rothemund's four tile types. Dark areas are able to bond to dark areas, light areas are able to bond to light areas.

tile shape and hydrophobic/hydrophillic wetting patterns on the edges of his tiles in order to encode the desired matching rules. For clarity, we'll work with the simplified version of Figure 6.10. The reader should bear in mind the fact that our tiles are symmetric, and we'll discuss them as if they remain oriented as shown in the figure. In Rothemund's experiments, this symmetry was broken by tile shape and orientation enforced by the broken symmetry. We've also avoided attempting to show Rothemund's wetting patterns. Instead, we've indicated via shading areas that bond differently. In our tiles, edges adjacent to shaded areas can bond to one another as can edges adjacent to white regions. Edges of opposite type do not bond. Notice in Figure 6.10 that each tile is labelled with a series of one's and zero's. The digits at the bottom of the tile indicate the type of binary string the tile can accept as an input. The digits at the top of the tile represent *two* copies of that tile's output. Further note that shaded areas encode ones while light areas encode zeros.

We can now imagine how tiles might compute on a bit string. Assume we placed our set of tiles in some suitable environment, allowing the capillary bond to operate properly. Further, assume we placed a target bit string in this environment, encoded in the same manner as our tiles. This situation is pictured in Figure 6.11. Ideally, tiles will now bind selectively to the input

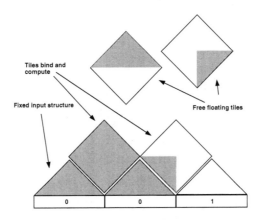

FIGURE 6.11: Computation on the bit string 001 using Rothemund's tiles.

structure as shown. The triangles encoding the string 00 present a gap and a binding rule that can only be filled by one of the tile types. This tile type is precisely the one that takes the binary input 00 and returns the appropriate XOR output, 0. Similarly, the triangles encoding the string 01 present a gap and a binding rule that can only be filled by the appropriate tile type.

Rothemund's experiments were quite a bit more extensive than described here. He not only set up a simple XOR operator, but allowed it to compute on a large bit string and effectively simulate the action of a one dimensional cellular automaton. Thus he not only computed with tiles, but used computation to build structures that were encoded in the tile types and input string. This is one of the great promises of this approach. By linking self-assembly with computation, we can begin to understand and define the structures that are accessible via a self-assembly process. The interested reader will find more details of these experiments in [106, 108].

6.3 Template Driven Self-Assembly

In Chapter 4, we introduced the idea of using a template to assist in the self-assembly process. The schematic of this process, Figure 4.4, was inspired by a set of experiments carried out by the Whitesides group and focused on template driven self-assembly [28].

In these experiments, the group extended their previous work with hexagonal particles and capillary interactions. This time, the particles and the templates were fabricated using photolithography and electrodeposition. This process allowed them to produce hexagonal particles with a side length of 10 microns and a thickness of 4 microns. Each hexagon was layered and consisted of a 4 micron thick gold middle layer, and two outer chromium thin film layers. The chromium layers were hydrophilic while the exposed gold surfaces were treated and made hydrophobic.

In addition to the particles, a variety of templates was also fabricated. These were made using the same process and had the same three layer structure as the hexagonal particles. The tops and bottoms of the templates were hydrophillic while the edges were treated and made hydrophobic. A typical template is shown in Figure 6.12. To carry out the assembly process, the

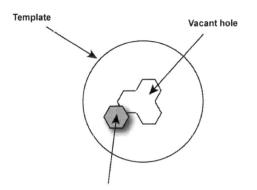

FIGURE 6.12: A typical template for the self-assembly of hexagons via capillary forces.

particles and templates were placed in a solution of nonpolar liquid adhesive. This adhesive selectively adhered to the hydrophobic faces of the particles and templates. The self-assembly process was now driven by a capillary force, similar to the solder dot system discussed above, as the liquid adhesive attempted to minimize its exposed surface area.

This work demonstrated the feasibility of using templates in a capillary bond based self-assembling system. It also revealed several difficulties that any such system presents. Experiments were conducted with templates designed to hold a single hexagon, three hexagons, seven hexagons, and nineteen hexagons. The group found that the yield of the experiments varied inversely with the size of the templates; the more hexagons needed to fill out a template, the lower the yield. The group also found that the process was prone to defects. Large assemblies tended to contain vacancies. Large and small templates occasionally produced assemblies with extra hexagons. Attempts were made to control the defects and to increase the yield of the experiments. Further details may be found in [28]. We also note that the Whitesides group has conducted at least two other studies using templates. In one, a solder dot type system was combined with a template to fabricate a cylindrical display system. In another, electrically charged surfaces were used as templates to construct a two dimensional pattern of gold disks. The reader is directed to [67] and [131] for details of these experiments.

6.3.1 Artificial Amphiphiles

The group led by Chad A. Mirkin of Northwestern University has also achieved great success with templates. Their group constructed artificial amphiphiles, like the lipids of Chapter 3, and showed how to force them to self-assemble into plates, cylinders, and spheres.

The rod-shaped amphiphiles of the Mirkin group consist of a hydrophillic gold "head group" and a hydrophobic conducting polymer "tail." These structures are true nanostructures, with a typical rod diameter of around 400 nanometers. The rod length is on order of 10 microns. Scanning electron microscope images of these rods are shown in Color Plate 11.9 (A). The process used to construct these rods involves the use of a template. The process begins with a porous aluminum structure that will serve as the template. Gold is deposited onto the template via electrodeposition. The polymer chains are attached to the gold head groups using an electrochemical polymerization method. Both the size of the gold head groups and the length of the polymer tails can be controlled. Further, in a given system, the rod-shaped amphiphiles can be made with great uniformity.

After the rods have been formed, the aluminum template is dissolved and the rods are allowed to self-assemble. Just as with the lipids of Chapter 2, these artificial amphiphiles organize themselves into structures in an attempt to satisfy the gold's water-loving desires and the polymer's water-hating desires. Also as with micelles, the structures that form can be characterized in terms of a geometric packing parameter. The Mirkin group's packing parameter can be written as

$$R = \frac{V}{a_h l_c} \tag{6.1}$$

where V is the volume of a rod, a_h is the average rod head area, and l_c is the rod length. The group found that increasing values of R corresponded with structures with less curvature. Intuitively, this is to be expected as increasing R implies decreasing a_h and hence the particles are becoming less cone-like and more rod-like in nature.

What is especially interesting about the Mirkin group experiments is the importance of the aluminum template. When dealing with lipids and self-assembling micelles, the lipids are simply placed in solution and micellar structures form. If however, the Mirkin group's artificial amphiphiles are simply placed in solution, they sediment out of that solution before they can self-assemble into two dimensional or three dimensional structures. It appears that the aluminum template orients the particles in such a way that self-assembly can occur before sedimentation takes over.

6.4   Structured Surfaces

We've encountered the idea of energy minimization several times in this text, and we've also discussed capillary forces that arise as surfaces attempt to minimize their free energy. But, thus far, we've avoided the rather obvious idea of using the minimization of surface energy to directly create structured surfaces. Such an approach brings us face to face with some of the richest areas in mathematics, namely the fields of minimal surfaces and constant mean curvature (CMC) surfaces.

The study of minimal and constant mean curvature surfaces dates back to the Belgian physicist Joseph Antoine Ferdinand Plateau who studied the properties of soap films in the late 1800's. His work, a mixture of theory and experiment, excited the interest of many of his mathematical contemporaries and continues to excite the interest of many present day mathematicians. The famous *Plateau Problem* in mathematics, which asks whether every closed curve in space can be spanned by at least one minimal surface, is an outgrowth of Plateau's experiments with soap films.

Recently, several groups have applied ideas from minimal and CMC surface theory, together with modern materials, to the fabrication of devices such as waveguides, microchannels, and microlenses. To better understand these experiments let's first examine a simple minimal surface and a simple constant mean curvature surface.

It's best to begin by following Plateau. Imagine taking two circular rings, aligned concentrically and held adjacent, dipping them in soap solution, and slowly pulling them apart. If you carried out this experiment, you'd form a surface between the rings similar to the one shown in Figure 6.13. This is a surface known as a *catenoid*. The soap film forming the catenoid attempts to minimize its surface area, or equivalently its surface energy. Using this

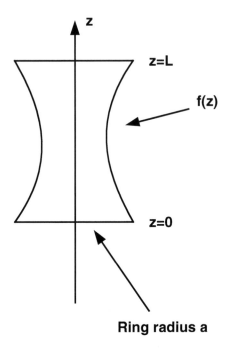

FIGURE 6.13: The geometry of the catenoid system.

fact and the idea of a functional, which was introduced in Chapter 2, we can compute the exact shape of the surface in Figure 6.13.

We assume that the surface of Figure 6.13 is a surface of revolution and hence may be described by a function of a single variable. In the coordinate system of Figure 6.13, we may write down the functional that describes the surface energy of our catenoid. We find

$$E[f(z)] = 2\pi T \int_0^L f(z)\sqrt{1 + f'^2}dz. \qquad (6.2)$$

Here, T denotes the tension in the soap film and the factor of 2π arises because of the fact that we are dealing with a surface of revolution. In addition to minimizing E, our soap film must also remain connected to the two rings and hence must satisfy the boundary conditions

$$f(0) = f(L) = a. \qquad (6.3)$$

As in Chapter 2, we are faced with the problem of minimizing a functional over some set of allowed functions. Fortunately, our functional is in integral form and we can apply the Euler-Lagrange equation. Applying the Euler-Lagrange equation to Equation (6.2) leads to a second order ordinary differential equation that may be easily integrated one time to yield

$$f\sqrt{1+f'^2} - \frac{f f'^2}{\sqrt{1+f'^2}} = c_0 \tag{6.4}$$

where c_0 is a constant of integration. This equation can be separated and integrated again to yield the shape of our surface. We find

$$f(z) = c_0 \cosh(\frac{z+c_1}{c_0}). \tag{6.5}$$

That is, the catenoid is simply a hyperbolic cosine rotated about an axis. The graph of the hyperbolic cosine is another shape that arises frequently in geometric problems; the shape known as the *catenary*.[1] To complete the description of our surface we must impose the boundary conditions and solve for c_0 and c_1. It is easy to see that $c_1 = L/2$, but that c_0 cannot easily be solved for explicitly. We find that c_0 satisfies

$$a = c_0 \cosh(\frac{L}{2c_0}). \tag{6.6}$$

To analyze this equation for c_0 it is useful to define $y = a/c_0$ and $\lambda = L/2a$. Equation (6.6) then becomes

$$y = \cosh(\lambda y). \tag{6.7}$$

Note that λ is a parameter in the problem. It measures the geometry. In particular it is the ratio of the ring radius to the gap between the rings. If we fix λ and solve for y, we can back out c_0 and hence obtain the shape of our surface. To see when this is possible it is useful to plot λ versus y from Equation (6.7). This plot is shown in Figure 6.14. Notice that for small values of λ there are two solutions for y. The lower branch of solutions corresponds to the catenoid that we see in experiment. The upper branch corresponds to a shape that maximizes the surface area. If λ becomes too large, no solution exists. This tells us that if the rings are pulled too far apart, we no longer have a catenoid spanning them. Further details of the analysis of the catenoid may be found in the books in the Related Reading section at the end of this chapter.

So, we are able to compute the shape of a soap film spanning two concentric rings. But, as we'll see, those using this approach to fabricate structures often introduce additional constraints into the system in order to allow a measure of control over surface shape. With the catenoid, we can control the shape by varying the gap between the rings, but unless we break the symmetry of the system, that is our only means of control. One method for extending the family of accessible shapes without breaking the symmetry of the geometry is to consider the case where the upper and lower rings are replaced by solid

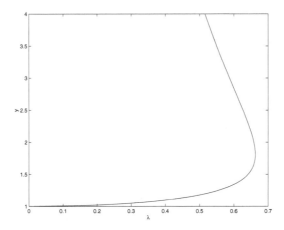

FIGURE 6.14: Solutions to the catenoid problem.

disks. This means that the shape spanning the disks now encloses a fixed volume. By poking a hole in one of the disks and forcing air into the system, we can vary this volume constraint. Mathematically, the addition of a volume constraint means that our surface must not only minimize E, but must also satisfy

$$V = \pi \int_0^L f^2(z)dz. \tag{6.8}$$

Here, V, is the fixed volume that the surface spanning the rings must enclose. To find the surface that minimizes $E[f(z)]$ subject to this constraint, we introduce a Lagrange multiplier, μ, and form the modified energy E^*.

$$E^*[f(z)] = 2\pi T \int_0^L f(z)\sqrt{1 + f'^2}dz + \mu\pi \int_0^L f^2(z)dz. \tag{6.9}$$

The Euler-Lagrange equation may now be applied to E^* in the same manner as above and a new differential equation for the surface shape derived. We leave this as an exercise for the reader. We note that this is now an example of a constant mean curvature surface rather than a minimal surface. The reader who completes the exercises for this section will see that this differential equation actually reduces to $H = 1$, where H is the mean curvature operator. Here however, we note two issues that arise when the analysis of the resulting differential equation is carried out. First, the differential equation is no longer easily solved in closed form. Numerical solutions or other techniques of analysis must be used. Second, it can be shown that this system exhibits the phenomenon of *symmetry breaking*. As the volume, V, is increased, the surface spanning the rings slowly bulges outward from the disks. At a critical value of the volume, the tangent to the function $f(z)$ becomes infinite at $z = 0$ and $z = a$. At this point, symmetry breaking occurs. In particular, the surface of least area that spans the disks and encloses the given volume is no longer a surface of revolution. Rather, the surface observed bulges out on one side and breaks the rotational symmetry.[2]

As mentioned above, several groups have applied the basic ideas behind minimal and CMC surfaces to the fabrication of actual devices. In a 1995 study, the Whitesides group, again working with polydimethylsiloxane (PDMS), fabricated a variety of structures using a setup similar to our soap film catenoid above [70]. In their study, the group placed droplets of PDMS between a pair of parallel plates. Since they were working with droplets the volume of their system was constrained. Further, the PDMS surface acted to minimize its area. Also, the group tailored the wetting properties of the two plates spanned by the PDMS. This meant, that unlike our soap films, the contact angle between the PDMS and the plates could be controlled. In the language of the problem above, the boundary conditions would change from specifying $f(0)$ and $f(L)$ to specifying $f'(0)$ and $f'(L)$.[3] Once a desired surface was formed, the PDMS could be cured and the surface solidified. In this way, the group was able to manufacture objects with various shapes. The team also explored other ways of controlling the morphology of surfaces. In particular, they could tailor the shape by changing the density of the surrounding liquid, applying electric fields, or by applying magnetic fields. Details of these experiments may be found in [70].

In another set of studies Lenz et al. tailored the wetting properties of surfaces, used vapor deposition to deposit water onto the surface, and in this way created microchannels [82, 81, 46]. In their system, which included ring shaped domains and straight channels, they observed a symmetry breaking phenomenon similar to the one described above. At a critical value of the amount of water vapor deposited onto a surface their previously symmetric channels lost symmetry and developed a bulge. An example of this is shown in Figure 6.15.

In this figure the liquid channels are constrained by the wetting properties of the surface on which they reside. The continued addition of water from the surrounding vapor represents a change in the volume constraint experienced by these surfaces. When a critical volume is reached, the surface that minimizes the energy of the system loses symmetry.

This approach to self-assembly puts a different twist on the forward, backward, and yield problems. The yield problem is avoided altogether; shapes are formed reliably. Some progress may be made on the forward problem. We showed how to find the shape between two rings. Extensions of this type have been explored by many researchers. However, complications such as symmetry breaking make even highly symmetric problems such as these difficult. When the geometry itself is not symmetric, the difficulty factor increases dramatically. Progress may often be made numerically and fortunately many computational tools have been developed to facilitate this. As far as the backward problem, well for this approach it is unfortunately all but intractable.

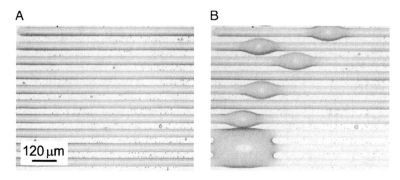

FIGURE 6.15: Microchannels of water constructed by Gau et al. Note the symmetry breaking bulge in the channels. From Gao, et al., Science, v. 283, pp. 46-49, (1999), Reprinted with permission from the AAAS.

6.5 Assembly by Folding

When we studied proteins in Chapter 3, we saw that the interesting aspect of their structure was formed by *folding*. In the *Try it Yourself* exercise in Chapter 3, we even saw a simple system, using the shape memory alloy Nitinol, that allowed one to explore how folding could lead to complex shapes. And, in this chapter, we saw geometric dissections solved via self-assembly of tethered particles, a different sort of folding. The folding motif, so powerful in natural systems, has also been explored by self-assembly researchers as a means for building engineered systems.

Probably the most complete and certainly the most interesting of these studies is the one carried out by Saul T. Griffith [54]. In this study Griffith designed and built two dimensional and three dimensional systems that self-assembled via deterministic folding. Here, we'll describe Griffith's two dimensional system. For details of the three dimensional system, the reader is referred to [54].

The first step in Griffith's approach was to develop a macro-scale mechanical analog of a protein. He required that his system be constrained to two dimensions, that it be composed of simple subunits, that those subunits be attached in a linear chain, and that the interactions between those subunits could be controlled. The simplest of his systems is the vertex connected linear string of square subunits shown in Figure 6.16 (a). The black dots in Figure 6.16 are pivot points. Squares must remain attached at these points, but are free to pivot about them. Griffith imagined chains being folded sequentially.

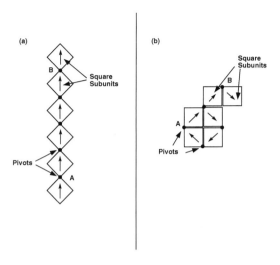

FIGURE 6.16:   The basic mechanical protein analog conceived by Griffith.

That is, we would start at the point labelled A in Figure 6.16 (a), fold about that pivot point, and then continue up the chain folding at each pivot in turn. Further, we can specify a folding sequence by giving a starting point and a direction in which to turn about each pivot. For example, the sequence of folds leading from Figure 6.16 (a) to Figure 6.16 (b) can be specified as follows. We start at the point labelled A and specify the direction of the fold about a pivot by the direction in which the arrowhead pointing to that pivot travels. We'll call a fold "left" or L if the direction is clockwise and "right" or R if the direction is counterclockwise. Hence, the folding sequence from Figure 6.16 (a) to Figure 6.16 (b) is given by $RRRLR$.

Now, given the labelling system and the vertex connected chain model, we can specify the folding sequence of many different shapes. But Griffith asked more of his system. First, he wanted to know if any shape in the plane composed of square tiles could be reached by folding his vertex connected chain. Second, he wanted to know whether binding forces could be assigned to the tiles in his chain in such a way that the chain would fold into a given shape on its own. That is, he wanted to know if a vertex connected chain could be made to self-assemble into a predefined pattern.

To answer the first of these questions, Griffith argued as follows. Imagine we tile the plane with square subunits as shown in Figure 6.17. Each arrow in the plane still points towards a vertex. The entire plane, or piece of the plane, can be covered with a single folded chain. In fact, this is a requirement of Griffith's model. Subunits other than squares can be used, but any chain must be able to fold into a tiling of the plane. Now, consider the two letters in Figure 6.17. A moments thought reveals that the letter "S" can be folded

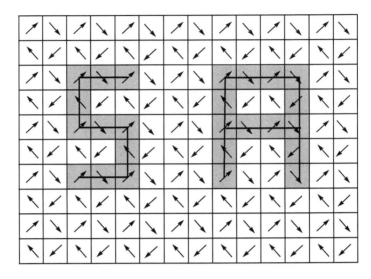

FIGURE 6.17: The construction of spanning trees for Griffith's vertex connected chain model.

from a vertex connected chain, but that the letter "A" cannot. To characterize those shapes that can be folded and those that cannot, Griffith reduced the problem to a problem in graph theory. Labelling the centers of the squares of the target object with nodes and defining a graph on this set of nodes, Griffith showed that a given object could be folded if and only if that graph contained an unbranched spanning tree. In essence, this is the difference between the line on the letter "S" and the line on the letter "A." By following each line we can visit the center of each square, but walking along the path on the letter "A" requires us to visit the same squares more than once. If the target object contains a path like the one on the letter "S," the object is foldable. If the only paths that can be found on a target object are branched paths like the one on the letter "A," the object is not foldable.

Griffith offered a way to construct objects that were not foldable by trading resolution for the ability to fold an object. He showed that if a given object could not be folded, we could fold an object with the same shape by dividing each of the squares of the original object into four squares. This fourfold increase in tiles is equivalent to a reduction in resolution. But, the new figure, with the same essential shape as the original, was foldable.

The next question Griffith addressed was whether or not shapes that were foldable could be made to fold themselves. To accomplish this he required that the notion of linear sequential folding be part of his self-assembly process. He imagined the chain being pushed through a nozzle into a large open space. Every time a new pivot passed through the nozzle, the fold about that pivot was to take place before the next pivot appeared. To force the folds to occur automatically, Griffith had to introduce a binding force into his system. Since Griffith intended to and eventually did build a working model of his system, he chose his binding force for simplicity of construction. In particular, he chose to work with magnets. The question then became one of how to place the magnets around the edges of the square subunits so that any folding sequence could be specified and so that the resulting bonds would be strong. He showed that by using four tile types, these goals could be met. Griffith's four tile types are shown in Figure 6.18. The labels "N" and "S" indicate

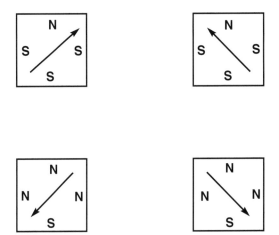

FIGURE 6.18:   Griffith's four tile types.

where the north and south faces of the four embedded magnets in each tile are facing outwards. The usual attractive and repulsive rules for magnets hold. Griffith's self-folding system works as described and is quite entertaining to watch. A link to his web page and video of the folding of the letters "MIT" is given in Appendix B and on the web page for this book.

The Griffith system nicely illustrates a method for solving the forward, backward, and yield problems of self-assembly. By forcing the system to use linear sequential folding, Griffith achieved effectively one hundred percent yield. The trade-off is in time and complexity. Since particle-particle interactions occur one at a time, the time of assembly is proportional to the length of the initial chain. Further, the chain does have to be fabricated and then forced through some sort of device like Griffith's nozzle. By reducing the problem from one of distinct particles to a system of particles along a chain, Griffith was also able to successfully attack the forward and backward problems for his system. If the binding forces, or particle types, are specified along the length of a folding chain, the sequence of folds is automatically determined and hence the final structure can be predicted. This amounts to solving the forward problem. The graph theoretic arguments demonstrating which shapes were foldable and which were not represents a solution of the backward problem. His method for turning unfoldable shapes into foldable ones with an accompanying loss in resolution also represents a solution to the backward problem. In short, this clever reduction of the general problem of self-assembly to the problem of the sequential folding of a chain provides one promising avenue for solving the forward, backward, and yield problems of self-assembly.

6.6 Chapter Highlights

- Static self-assembling systems are those that function via a principle of energy minimization. The initially disordered system moves down an energy gradient towards an ordered equilibrium configuration.

- Systems using the capillary bond as their binding force have been explored by several groups. The forward, backward, and yield problems of self-assembly arise in these systems in a transparent manner.

- Capillary based systems have been pushed beyond "mere crystallization." In particular, they have been shown to be reconfigurable and to allow for computation.

- Templates can also be used in static self-assembly. These can reduce the complexity of the backward problem or can serve as an intermediate step in an assembly process.

- Direct minimization of surface energy provides a method for assembling structured surfaces. By introducing a variety of novel constraints into the system, structures beyond typical "soap-film" structures can be formed.

- Self-assembly by folding of a connected linear chain allows the general problem of self-assembly to be reduced to one where the forward, backward, and yield problems can be attacked.

6.7   Exercises

Section 6.2

1. Consider the solder dot problem presented in this section. Sketch the energy surface for squares having a single central dot and allowed to move in the plane rather than restricted to motion along a line. Do the same for squares with two dots. How do these landscapes differ?

2. Attempt to solve the backward solder dot problem for square faces needing two dots. Enforce the condition that your dots must be some minimum distance apart. Sketch the energy landscape for your design assuming the squares are restricted to move in the plane. How successful is your design? Can you abstract any design rules that might apply to the three dot problem? How about the n dot problem?

3. Using the simplified version of Rothemund's tiles presented here, simulate the assembly process using the input string ...00100... where dots indicate a string of zeros continuing on both sides. Follow the process through multiple layers. That is, once the initial string is filled in, allow new tiles to compute on the output string. What structure emerges?

4. Consider other logical operators such as AND and the inclusive OR operator. Design tile types that allow the AND and OR operation to be performed on an input bit string.

Section 6.3

5. The use of electrostatic templates was mentioned in this chapter. In particular, the reader is directed to [131]. Read this article and explain how this template differs from the other templates discussed in this chapter.

6. Yet another sort of template was explored by Lopez et al. in [87]. This one involved the condensation of liquid from a vapor onto a patterned surface. Read this article and explain how this template was constructed. Also explain how this template, and the structures formed, are related to the capillary surfaces of Section 6.2.

Section 6.4

7. Carry out the minimization of the modified functional of this section to derive the equation that governs the shape of the volume constrained system. Show that for a proper choice of the volume, $f(z) = a$ satisfies the equation and the boundary conditions.

8. Show that the differential equation you derived in the last exercise can be written as $H = 1$, where H is the mean curvature of the surface spanning the disks.

Section 6.5

9. The linear folding sequence leading to a given shape for a vertex connected string is not in general unique. Find all of the folding sequences that you can that lead from Figure 6.16 (a) to Figure 6.16 (b).

10. Griffith's linear folding model need not be carried out using squares. However, Griffith required that the subunits be able to tile the plane. Explain why and give several examples of allowable subunits other than squares.

11. Consider the letter "A" in Figure 6.17. Show that if each square is divided into four smaller squares, the resulting structure can be folded from a vertex connected chain of squares.

6.8   Related Reading

In creating a physical system to solve their geometric dissection problem, the Whitesides group actually created a *hinged dissection*. Very little has been done with regards to designing self-reconfiguring hinged dissections. The interested reader will find the book by Frederickson invaluable.

G.N. Frederickson, *Hinged Dissections: Swinging and Twisting*, Cambridge: Cambridge University Press, 2002.

A very entertaining look at computation divorced from silicon is the book by Dewdney.

A.K. Dewdney, *The Tinkertoy Computer and other machinations*, New York: W.H. Freeman and Company, 1993.

The books by Finn, Oprea, and Isenberg have already been mentioned. However, they are also useful resources for the reader interested in minimal and capillary surfaces.

R. Finn, *Equilibrium Capillary Surfaces*, New York: Springer-Verlag, 1986.

J. Oprea, *The Mathematics of Soap Films: Explorations with Maple*, American Mathematical Society, 2000.

C. Isenberg, *The Science of Soap Films and Soap Bubbles*, New York: Dover, 1992.

The beautifully illustrated book by Hildebrandt and Tromba recounts much of the history of the study of minimal and capillary surfaces.

S. Hildebrandt and A. Tromba, *The Parsimonious Universe: Shape and Form in the Natural World*, New York: Springer-Verlag, 1996.

6.9   Notes

1. The catenary is, of course, the shape assumed by a chain hanging in a gravitational field.

2. The reader should consult the works of H. Wente, R. Finn, and S. Hildebrandt for a complete analysis of symmetry breaking in capillary surfaces. The mathematical analysis of such problems often becomes subtle and difficult.

3. This type of surface is also called a liquid bridge.

Chapter 7

Dynamic Self-Assembly

Science moves, but slowly, slowly, creeping on from point to point.

Tennyson, *Locksley Hall*

7.1 Introduction

In Chapter 1 we examined the reasons why there is so much interest in self-assembly today. We offered a quote from Richard P. Feynman that captured one reason for much of the excitement:

> The biological example of writing information on a small scale has inspired me to think of something that should be possible. Biology is not simply writing information; it is *doing something* about it. A biological system can be exceedingly small. Many of the cells are very tiny, but they are very active; they manufacture various substances; they walk around; they wiggle; and they do all kinds of marvellous things – all on a very small scale. Also, they store information. Consider the possibility that we too can make a thing very small which does what we want – that we can manufacture an object that maneuvers at that level!

In the last six chapters we've strayed a bit from Feynman's vision. None of the systems that we've described thus far are terribly active. They don't manufacture, they don't walk around, they don't wiggle; they might be marvellous in their own right, but, they aren't very active. As we saw in the last chapter, most of the systems we've been studying are examples of *static* self-assembling systems. In this chapter we turn our attention to systems that do walk around, do wiggle, and do maneuver. We examine *dynamic* self-assembling systems.

In Chapter 1, we defined dynamic self-assembly. We categorized dynamic self-assembling systems as a subclass of self-assembling systems that leads to stable non-equilibrium structures. We said that these structures remained ordered only so long as the system continued to dissipate energy. The inspiration for studying such systems is precisely that identified by Feynman – biology.

Although the biological examples we studied in Chapter 3 were largely static, it does not take much imagination to see that most of biology is dynamic and must rely upon some form of dynamic self-assembly. If only we could learn to do what nature does so easily, build cells, with all the concomitant properties that implies, we'd have achieved the ultimate goal in self-assembly research.

Sadly, we're not yet close. Research in dynamic self-assembly is truly in its infancy. Yet, some progress has been made, and that is the subject of this chapter. Here, we'll examine what has been done, attempt to understand dynamic self-assembly from the perspective of Chapter 4, and attempt to understand the sorts of unique design problem that arise in this area.

We begin in Section 7.2 with a detailed look at a prototypical example of dynamic self-assembly. Our system consists of conducting particles suspended in a dielectric fluid and placed in an electric field. As we'll see, these particles become active and form simple ordered structures. We'll carefully look at this system from the viewpoint of Chapter 4 and identify the role of structured particles, binding forces, an environment, and driving forces in this system. We'll learn that one feature of dynamic self-assembling systems is that particle interaction must be competitive. At times, particles attract, at other times the same particles repel. In addition, this interaction can depend on externally applied environmental conditions. After discussing our simple prototype system, we'll examine several examples of engineered systems that use conducting particles in an electric field to create nanoscale and microscale structures. We'll see how this idea can be applied to the creation of nanoscale wires only a few tens of nanometers in diameter, but several microns in length. We'll also see how the properties of the prototype translate into a *self-healing* behavior of the nanoscale system. Next, we'll see how driving a system of particles using an electric field can produce complicated temporal and spatial patterns in two and three dimensions. Again, these structures appear only while the system is dissipating energy, and they can be controlled by manipulating the applied field. To conclude this section we'll take a brief look at *electrorheological fluids*. The electrorheological system is essentially the same as the other systems of this section, but the number of particles has been scaled up massively. We'll see how under the action of an applied field interesting net-like structures can appear in these fluids. We'll also see that from a macroscopic point of view creating and manipulating these structures can allow us to bring about dramatic changes in the bulk behavior of the fluid.

In Section 7.3 we examine magnetically driven dynamic self-assembling systems. At first glance, these systems strongly resemble the magnetic self-assembling systems described in Chapter 5. Here however, in addition to magnetic forces the system also relies on the presence of a repulsive hydrodynamic force. In the systems of Chapter 5, magnetic forces, either attractive or repulsive, locked the particles in place. Here, the magnetic force not only attracts the particle towards the center of the system but sets the particles in motion. In turn, this creates a flow in a surrounding fluid. This flow creates the repulsive hydrodynamic force between particles. As we'll see, this competition between forces creates interesting dynamic structures.

In Section 7.4 we'll take a brief look at two mechanically driven dynamic self-assembling systems. The design of these experiments is very similar to the design of the electrostatic self-assembling systems of Section 7.2, but of course, there is no electrostatic force. Nonetheless, these systems do produce complex spatial and temporal structures in two and three dimensions. The first of these two systems accomplishes this via the use of hydrodynamic forces that are both attractive and repulsive. The second of these systems is not yet understood. However, experimentalists have been able to map out a phase space for this system that shown a rich and unexpected set of behaviors.

In Section 7.5 we look at one system that accomplishes dynamic self-assembly without the use of an externally applied field. In this system, particles interact with their environment to produce a driving force. The particles are capable of binding to one another and do so as they swim about the surface of a fluid. Once bound, they continue moving, creating larger, mobile, structured complexes.

In the final section of this chapter, Section 7.6, we ask the question - What if our particles were smart? That is, if instead of merely reacting to their environment and to the presence of other particles, we imagine what might be possible if our particles could make decisions for themselves. We explain how smart particles would allow us to more successfully attack problems like the yield problem of self-assembly. We see that for a particle to be considered truly intelligent, it must be able to sense its state, communicate with its neighbors, and act on this information. We conclude with a look at a dynamic self-assembling system that uses smart particles to carry out programmable self-assembly.

7.2 A Prototype for Dynamic Self-Assembly

Take an ordinary rubber balloon, rub it on your head,[1] and you'll find it sticks to ceilings and walls. Unless your head is covered with glue, the force that allows the balloon to stick is the *Coulomb force* or more simply the *electrostatic force*. If you rub two balloons on your head, charging them in the same way, and then attempt to bring them together you'll find that they repel one another. The electrostatic force can be attractive or repulsive.

It is the combination of these attractive and repulsive properties of the electrostatic force that allows our prototypical dynamic self-assembling system to function and create ordered structures. The basic setup of the system we'll consider is shown in Figure 7.1. The system[2] consists of a small plastic tray about 15cm long, 10cm wide, and 2.5cm deep that is partially filled with a dielectric fluid. The dielectric constant of the fluid[3] used in the experiments

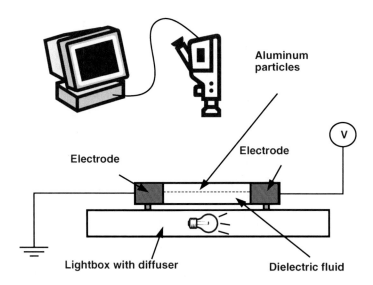

Aluminum particles

Electrode

Electrode

V

Lightbox with diffuser

Dielectric fluid

FIGURE 7.1:   The experimental setup for our prototypical example of dynamic self-assembly.

described here was approximately 2.6. The fluid covers the tray to a depth of about 1cm. The tray also holds a pair of large aluminum blocks that serve as electrodes. Suspended in the fluid are several small, thin, aluminum disks punched from a sheet of foil. The electrodes are attached to a regulated high voltage power supply capable of producing a potential difference of up to 100kV across the electrodes. The tray sits on top of a photographic light box. This enhances the contrast for the video camera positioned above the system. The video camera is attached to a video monitor allowing for easy observation during the experiments.

So, we have the basic ingredients, but what does this system do? Well, in a typical experiment the aluminum particles are scattered randomly in the fluid and then the voltage is turned on. At the outset, the particles are distributed like those in Figure 7.2 at time $t = 0$ or in Figure 7.3 again at time $t = 0$. Once the voltage is applied, small charge imbalances on the particles cause them to drift towards one of the electrodes. Some particles will contain a slight negative charge, others will contain a slight positive charge. The particles will migrate towards the electrode that contains the opposite charge. Now, when a particle reaches an electrode it gives up its charge to that electrode and acquires the *same* charge as the electrode. But this means that the particle will be now be repelled from that electrode. Such a particle accelerates *away* from the electrode it was previously attracted towards and moves to make contact with the opposite electrode. When it reaches that electrode the process repeats. If we placed but a single particle in the fluid we'd simply

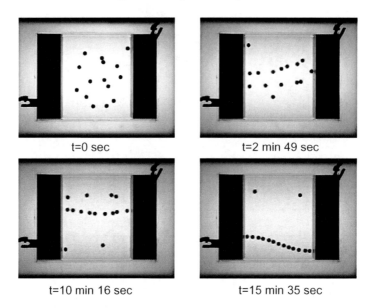

t=0 sec   t=2 min 49 sec

t=10 min 16 sec   t=15 min 35 sec

FIGURE 7.2: Frames extracted from video of our prototypical example of dynamic self-assembly. This experiment used many small particles. The applied voltage was approximately 19.8kV. Photograph by the author/MEC Lab - University of Delaware.

see it bounce back and forth between the electrodes.[4] But, there are other particles in the fluid trying to behave in the same way. As these particles pass each other, they interact. If a particle passes another particle carrying the opposite charge both particles experience an attractive force towards each other. If they encounter a particle with the same charge, they experience a repulsive force forcing them away from one another. Remarkably, over time, these repeated interactions cause the particles to self-assemble into a linear structure. This evolution can be seen in Figures 7.2 and 7.3. Notice how the vertical spread of the particles gradually narrows over time until a simple chain is formed. Particles not in the chain eventually join with little disruption to the overall dynamic behavior. In experiments the chain itself may drift, but this occurs on a timescale much longer than the timescale on which the particles oscillate within the chain.

Once the chain has formed, it persists. The system is dynamically self-assembling. If the electric field is suddenly turned off, the particles will slowly stop moving as the charge in the system is bled off, eventually returning to their initial disordered state. While in a chain each particle undergoes oscillatory motion. The particles move back and forth in a regular manner alternately colliding with and exchanging charge with their neighbors.

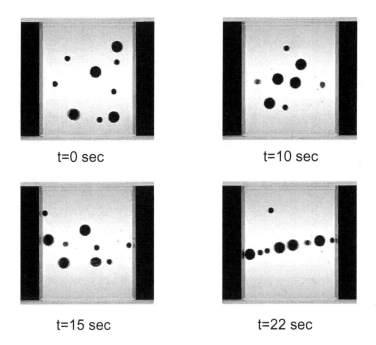

t=0 sec   t=10 sec

t=15 sec   t=22 sec

FIGURE 7.3: Frames extracted from video of our prototypical example of dynamic self-assembly. This experiment used a mixture of large and small particles. The applied voltage was approximately 19.8kV. Photograph by the author/MEC Lab - University of Delaware.

This system may be modified in several ways. Many of these lead to either changes in the approach to the final structure or to changes in the final structure itself. Notice that in Figure 7.3 a mixture of particles with different diameters is used. At the same applied voltage and same gap between the electrodes, the system with the mixture forms a chain much more rapidly than the system using only small particles. From the figures we see that full chain formation took more than 15 minutes in the small particle system, but less than a minute in the system with mixed particle diameters. Something not apparent in the figures is that the dynamics of the particles in the assembled chain also varies between the mixed and nonmixed systems. In the system with a uniform particle distribution, particles in the chain all oscillate with roughly the same frequency. In the system with large and small disks, the small particles oscillate with a frequency much higher than the large particles. Other system variables may be changed. For example, the disks may be replaced by squares, rectangles, or any shape we choose. If rectangular particles are used, they are found to form a chain and in addition orient themselves along the field lines between the electrodes. If more structure is added to the

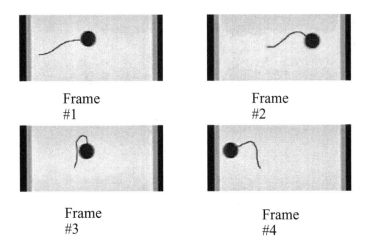

Frame
#1

Frame
#2

Frame
#3

Frame
#4

FIGURE 7.4: Frames extracted from video of the motion of a structured particle in our prototype system. Note, this particle always moves "head first." Photograph by the author/MEC Lab - University of Delaware.

particles, the changes can be even more dramatic. In Figure 7.4 we see an example of the motion of a particle consisting of an aluminum disk attached to a flexible tail. In this case, the motion of the particle is oriented. It always moves "head first," keeping the tail behind and the disk pointed towards the electrode to which it is travelling.

This simple system fits our definition of a dynamic self-assembling system. When the electric field is not present, the system remains in a state of disorder. When the field is applied the system rapidly organizes itself into a stable structure. In this state the system is dissipating energy. Some energy is lost as heat due to resistance in the electrodes and the particles while some energy is dissipated by the particles in the fluid. Yet, this system is comprised of the same four key components of self-assembling systems we identified in Chapter 4. Let's examine each of these components more closely in the context of this system and attempt to identify how this dynamic system differs from the numerous static systems we've seen.

We'll take each component in turn. In this system, the aluminum disks play the role of *structured particles*. At first glance, there is very little structure to the particles. They are simple disks. However, structure means more than simply the physical shape. As we saw in the last chapter, structure can come from a wetting pattern along particle edges or from a pattern of magnets embedded within the particle. Here, the additional structure comes from the fact that the particles are conducting. Each particle can hold a charge, so to specify the state of a particle we must give both its shape and its charge. The charge on a particle effectively represents the particles conformation.

Try It Yourself - Electrostatic Self-Assembly

In this chapter we've seen how electrostatic forces applied to suspended systems of particles can produce complex and interesting structures. It is possible to carry out some of these experiments with easily obtainable materials. The only specialized equipment you'll need is a high voltage source. However, an ordinary Van de Graaff generator works fine.

Materials

- High voltage source

- Corn oil (The cheaper the brand the better it seems to work.)

- Rice

- Aluminum foil

- A clear plastic container (A container the size and shape of an ordinary drinking glass is perfect.)

- Wire

- Glue

Procedure Drill a small hole in the bottom of your container. Thread a piece of wire through the hole and laminate a layer of foil onto the bottom of the container from the *inside*. Make sure the foil makes contact with the wire. This will serve as your lower electrode. Fashion a second electrode from a piece of foil and attach a wire to this electrode. This electrode will float on the surface of the corn oil. Now, partially fill your container with corn oil. The depth will depend on the strength of your high voltage source. Start with 2cm. You can add more oil if needed. Sprinkle a handful of rice grains into the corn oil. Allow the rice to sink. Now, float your second electrode on top of the corn oil. Attach your high voltage power supply and turn it on. You should see the rice grains begin to move and form chains as in the systems described in this chapter.

Things to Try

- Let your system self-assemble a chain and then switch off the high voltage source. How long does it take for the chain to collapse?

- Try using very few rice grains, fewer than are needed to bridge the gap between your electrodes. What happens when the high voltage source is switched on?

Further Reading If you don't have access to a Van de Graaff generator, you can always build a *Dirod*. The construction of this clever electrostatic generator is described in *Electrostatics: Exploring, Controlling and Using Static Electricity* by A.D. Moore and J.M. Crowley.

The *binding force* in this system is the electrostatic force. Analogous to capillary systems where particles could be hydrophobic or hydrophillic, here particles can be positively charged or negatively charged. If a pair of particles is positively charged they repel, if they are both negatively charged they repel, but if they hold opposite charges, they feel an attractive force. However, unlike capillary bond systems, where the wetting properties of the particles remain fixed, here, the particles can change their type as they interact with the environment and with each other. We've already seen that contact with an electrode causes a particle to flip from one charged state to another. If particles themselves make contact this flipping can also occur. When the particles are organized into a chain they do this repeatedly, moving back and forth, making contact with their neighbors, and changing their state.

The *environment* in this system includes the fluid in which the particles are immersed, the gravitational field in which the system sits, and really the presence of the external applied electric field. We'll classify the applied electric field as a *driving force* and hold off on discussing it further for a moment. The fluid in which the particles reside exerts a drag force on the particles. Because of the scale, in this system drag forces are quite significant and inertial forces all but negligible. This means that the particles are observed to move with constant velocity rather than constant acceleration. Without the fluid, a particle leaving an electrode would accelerate continuously towards the opposite electrode. With the fluid, this acceleration is rapidly balanced by the drag force and the particles cross most of the gap with constant velocity. The gravitational field serves to orient the particles. Note that in Figures 7.2 through 7.4 every single disk is oriented with its largest side facing the camera. This is not because the particles do not have room to rotate; the particle's largest length scale is still small compared to the depth of the fluid. This is because the combination of gravity and lift forces generated on the particles as they move through the fluid keeps them oriented.

Finally, the *driving force* in this system is the applied electric field. In the absence of an applied field the particles sit in a disordered state. Perhaps, given enough time, small charge imbalance on the particles will force them to aggregate and form an ordered crystalline structure like those considered in Chapter 6. However, they will not form a chain, and will not exhibit the back and forth oscillatory motion that occurs when the field is applied. Further, the presence of the electric field and the contact with the two electrodes causes the particles to acquire a charge much larger than any they might have had initially. Further contact with the electrodes allows the particles to switch their charged states, changing their conformation.

We see that the key components of a self-assembling system identified in Chapter 4 are all present in our prototypical dynamic self-assembling system. Yet, the interaction of these components is more complex than in the static systems we've considered thus far. The major differences between this system and the static systems of previous chapters lie both in the nature of particle-particle interactions and in the interaction of particles with the environment.

Particles in this system do not remain in a single state. Rather, they continuously switch their states thereby continually modifying their interactions with neighboring particles. Further, the applied electric field supplies the energy source for this system and causes the particles to change their state. We should bear in mind that this system is constantly dissipating energy. Unlike static systems, when the energy supplying the driving force is removed from this system, the self-assembled structure disappears.

In designing a dynamic self-assembling structure such as this one, the self-assembly researcher is still faced with the forward, backward, and yield problems identified in Chapter 5. Here, the forward problem is as straightforward as it was for many of the systems in Chapter 6. Given the setup of the system, predict what the particles will do. Details of particle interactions and the nature of the environment and driving force may make this difficult, but the question is posed easily enough and is still relevant. The yield problem also appears. We could ask, for example, what percentage of the particles initially distributed in the fluid will end up in our ordered chain. We see from Figures 7.2 and 7.3 that the yield is not one hundred percent. Here, it may be a question of time. How long must we wait to ensure that some suitable percentage of our particles has joined a chain? As usual, the backward problem is present and is the most difficult to attack. We could ask: How might we tailor our particle shapes, our choice of fluid, and the geometry of the applied field to form a given dynamic structure? In the remainder of this section we'll explore various systems that demonstrate the range of possible answers to this question.

7.2.1 Self-Assembling Nanowires

In the system described above the sum of the diameters of all of the particles used was less than the gap between the electrodes; the particles could not span the gap. This caused the system to enter into a dynamic ordered state where the particles oscillated continuously. In the systems we'll consider here, many more particles are used and linear structures formed can span gaps between electrodes. This allows one to self-assemble wires. The wires are still dynamic structures but instead of an oscillating dynamic chain, they form a continuous conducting wire.

In an attempt to create nanoscale wires with novel electrical properties Bezryadin et al. [13] experimented with graphitized carbon nanoparticles in an electric field. A scanning electron microscope image of their assembled wires is shown in Figure 7.5. To create these wires the group dispersed graphitized carbon nanoparticles, approximately 30 nanometers in diameter, in the dielectric fluid toulene. The group then used standard lithography techniques to fabricate a pair of chromium electrodes on the surface of a piece of silicon. The electrodes were about 10 nanometers thick and placed about 1 micron apart. The silicon wafer was then immersed in the toulene solution and a potential difference was applied across the chromium electrodes. The applied

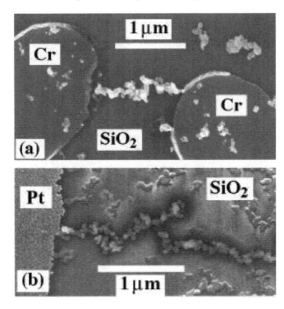

FIGURE 7.5: Scanning electron microscope images of a self-assembled chain of carbon nanoparticles. Reprinted with permission from Bezryadin et al., Applied Physics Letters, v. 74, 1999, pp. 2699. Copyright 1999 by the American Institute of Physics.

potential was 40 volts. In addition, the group placed a large 1 giga-Ohm resistor in series with the voltage source. The presence of the resistor ensured that only one wire formed.

As the system evolved Bezryadin et al. monitored the current. The initially very small current through the system jumped by several orders of magnitude after only a few seconds. In this way, the group knew that a continuous wire had been formed. Since the group could not directly observe their wires during the assembly process they immediately removed their silicon plate from the solution when they observed the current jump. The plate could then be dried and the wire structure preserved. This then allowed them to obtain SEM images of their assembled wires such as the ones shown in Figure 7.5. During their experiments the group managed to grow nanowires as long as 6 microns in length. The bottom half of Figure 7.5 shows a portion of their longest chain. Note that the aspect ratio of their longest wire, the ratio of length to diameter, is approximately 200, 000. For comparison, the aspect ratio of a piece of human hair 6 microns long is only about 1/10.

Now, it is important to note that while the basic features of the Bezryadin et al. experiments are the same as those of the large scale system described above, due to scale effects the dominant forces in the assembly process likely differ. Perhaps more important to note is that unlike the toy system above, the nanowire system of Bezryadin et al. promises to be directly useful in applications. As mentioned above, one of the primary goals of their work was to fabricate nanowires with novel electrical properties. This possibility arises because the system operates at the nanoscale. At this scale, quantum effects come into play and the charge transport from particle to particle is affected. In particular, in such systems, the well known Coulomb blockade occurs. A variety of devices such as the single electron transistor have been designed on the basis of this phenomenon. The reader is referred to [13] for more details concerning the possible application of nanowires.

In a related set of experiments, Hermanson et al. [61] fabricated microscale wires from suspensions of nanoparticles. However, this assembly process was carried out using AC applied voltages and explicitly relied upon the phenomenon of *dielectrophoresis* (DEP). In DEP particles move because they become polarized by an applied electric field. When a dielectric particle is placed in an electric field there is a nonzero field within the particle. This is in contrast to a perfectly conducting particle where the internal electric field is identically zero. The presence of an electric field in the dielectric particle causes charge migration within the particle, resulting in polarization. While the particle is overall charge neutral, the distribution of charge is such that one end of the particle appears negatively charged and the other positively charged. Two examples of this appear in Figure 7.6. The dielectric constant

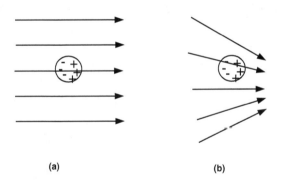

(a)   (b)

FIGURE 7.6: The behavior of a dielectric particle in a uniform electric field (a) and a converging electric field (b).

of the material is in fact a measure of the extent to which the material becomes polarized in this situation. If the surrounding field is spatially uniform forces on these charges exactly balance and the particle feels no net force as in the left hand sketch of Figure 7.6. On the other hand, if the surrounding field is spatially nonuniform, forces on these charges no longer balance, and the particle feels a net force as in the right hand sketch of Figure 7.6. In simplified situations these forces may be computed exactly. The reader is referred to [101] for further discussion of DEP and a derivation of the exact forces on a dielectric particle with spherical shape.

Here, our main interest is in how Hermanson et al. used DEP in the design of a dynamic self-assembling system. In constructing their system this group used gold nanoparticles between 15 and 30 nanometers in diameter. These particles were suspended in water and the mixture was placed between a pair of planar metal electrodes. The gap between the electrodes could be varied from microns to centimeters in length. The applied alternating potential in these experiments was between 50 and 250 volts, alternating with a frequency of 50 to 200Hz. The results of several of these experiments can be seen in Figure 7.7. In this case, the group could observe the dynamic assembly

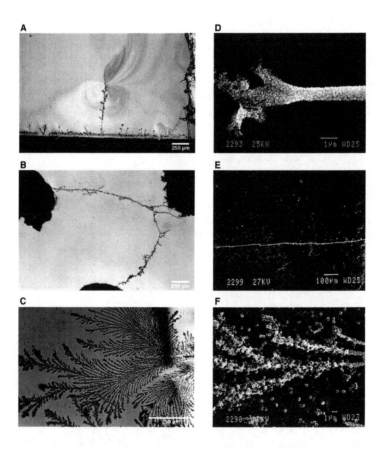

FIGURE 7.7: Images of growing nanoparticle wires from Hermanson et al. From Hermanson, et al., Science, v. 294, pp. 1082-1086, (2001), Reprinted with permission from the AAAS.

process. When the voltage was applied they observed thin fibers growing from the electrode towards the other electrode. The fibers grew at approximately 50 microns per second. The group conjectured that the fiber growth was due to a combination of the DEP effect and complex hydrodynamic effects occurring at the end of the growing wires.

Hermanson et al. were able to achieve a good deal of control over the structure of their growing fibers by varying several environmental parameters. They studied the effects of variations in the applied voltage, the AC frequency, particle concentration, particle size, and electrolyte concentration in their solution. They found that under the right conditions they could assemble branched wires and systems of interconnects spanning multiple electrodes. Further, they were able to assemble complicated dendritic structures such as the one shown in Figure 7.7 (c). The group also noted that their structures were *self-healing*. They demonstrated this by increasing the current through an assembled microwire until the wire snapped like a fuse. Quickly, new particles moved in to fill the gap and the wire reassembled.

In yet another set of experiments demonstrating self healing of a wire, Dueweke et al. [35, 127] worked with a system similar to our prototype, but in a different geometry. Their system consisted of a layer of dielectric fluid in a petri dish. The diameter of the dish was 140 millimeters and the fluid layer was 5 millimeters thick. Metallic spherical particles about 1 millimeter in diameter were deposited in the fluid. The group then inserted a pair of point electrodes into the dish and applied a potential difference between the electrodes. The potential used was between 15kV and 25kV. As in the experiments discussed above, the particles moved and formed a chain bridging the gap between the electrodes. Also as above, if the chain was disturbed the system reassembled and repaired the chain.

What makes the work of Dueweke et al. so interesting is not only the results of their experiments but the fact that they posed a variational principle that appears to govern the dynamics of chain growth. One of the difficulties in modelling dynamic self-assembling systems is the lack of such variational principles. In contrast to static systems, where an energy minimization principle can usually be uncovered, in dynamic system energy is not minimized. Duewek et al. proposed that their system minimized *resistance*. They defined the total resistance of their system as the ratio of the applied potential difference between their electrodes to the total current flowing between these electrodes. Then, neglecting inertial forces, they introduced an equation of motion for each particle in the system. In particular,

$$\gamma \vec{r}_i = \frac{1}{2\epsilon} \int_S \vec{E}_i^2 da_i \vec{n}_i. \qquad (7.1)$$

Here, $\vec{r}_i$ is the position vector of the ith sphere, ϵ is the permittivity of the fluid, γ is a drag coefficient, and $\vec{E}_i$ is the electric field at the surface of the ith particle. The integration is performed over the surface of the sphere. Next, using the method of images and the principle of superposition, they were able to eliminate the field from the right hand side of their equation of motion in favor of a complicated expression only dependent upon the $\vec{r}_i$. In turn, this allowed them to compute the time rate of change of the total resistance in their system. They found that the total resistance was in fact a Lyapunov function for their dynamic equations. That is, the total resistance acted like an energy function for their system and could be shown to decrease with time. At least in the context of this model, their principle of minimum resistance held.

7.2.2 Electrostatically Driven Granular Media

The group led by Igor Aranson has conducted numerous experiments with a system similar to our prototype, but flipped on its side so that particles must also struggle against gravity [6, 7, 116]. Their basic setup is shown in Figure 7.8. Aranson et al. worked with a variety of different experimental

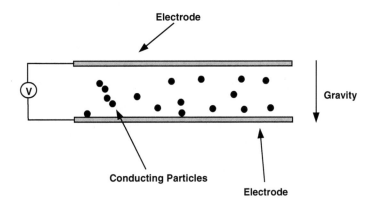

FIGURE 7.8: The basic experimental setup for electrostatic self-assembly when the electric field competes with gravity.

setups, we'll discuss the one described in [6]. In this experiment, 4cm by 6cm plate electrodes were used. The top plate was transparent so that images could be captured from above. The gap between the plates was set at 1.5mm and filled with either air or vacuum. Roughly ten million spherical copper particles 35 microns in diameter were placed in the gap. The team studied both the effects of DC and AC voltage on their system. Voltages of several kilovolts were applied with a frequency ranging from zero to 250Hz.

In this system, when the voltage is first applied all of the particles reside on the lower plate. Since they are in contact with the lower electrode they immediately acquire a charge. The magnitude of this charge depends on the strength of the applied field. Here, in order to move, the particles must overcome the force of gravity, which acts to oppose their initial motion. This does not occur until a critical value of the applied field is reached. When a particle does move upward and make contact with the upper plate, it deposits its charge and then falls back towards the lower plate. By applying an AC voltage the group found that the height reached by the particles could be controlled. The switching of the electric field allowed them to push the particles back towards the ground before they made contact with the upper plate.

Mapping out the behavior of the system as a function of the applied voltage and frequency the group found three distinct regimes of behavior. In the first regime, no motion occurred. The threshold value of the applied voltage was not yet reached. If the applied voltage was too large the system entered into a phase where the spheres behaved like a granular gas. In this phase, the sphere are dispersed uniformly. The middle or second phase showed the most interesting behavior. The group called this phase the *coarsening* phase. In this state, as particles moved up and down, they also clustered in the horizontal direction. Viewed from above one sees large dots begin to appear. Over time these dots grow and grow together. When examined closely one sees that the particles in these dots are in fact stationary. The particles have formed a thick chain spanning the gap. Remarkably, over time, these dots coalesce and form one large almost perfectly circular dot. Also within the coarsening regime, the group demonstrated that both the number of clusters and the average cluster radius showed a power law dependence on time. In particular, average cluster radius grew like $\sqrt{t}$ and the number of clusters decreased like $1/t$.

As mentioned above, the Aranson group conducted multiple experiments with the setup described here as well as with variations of this system. Using fluid filled cells the group constructed a system with an even richer phase diagram. This system formed honeycomb shaped structures, torus shaped vortices and pulsating rings. To truly appreciate these structures the video is essential. A link may be found in Appendix B.

The dynamic self-assembling system of Aranson et al. nicely illustrates the fact that a range of interesting behaviors can often be observed in a single system. As is typical, they characterized this range by constructing a phase diagram for their system. In so doing, they demonstrated how fine control over the parameters in the system can translate into fine control over the behavior of the system. This represents a partial attack on the backward problem of self-assembly. Given the characterization of the structures that the system can produce, Aranson et al. showed how to design the system, by tuning of the parameters, to select for a particular structure.

Finally, we note that many other groups have studied systems similar to those described here. The reader is referred to [33, 94] for descriptions of two other particularly interesting studies.

7.2.3 Electrorheological Fluids

If the number of particles in the systems described above is increased by several orders of magnitude, we obtain what is known as an *electrorheological fluid*. As with the ferrofluids described in Chapter 5, the focus of studies with electrorheological (ER) fluids is usually on bulk changes to the rheology of the fluid as opposed to dynamic structures formed by the particles in the fluid. Nonetheless, because of their great potential for applications and their relationship to the systems we've been discussing, we briefly describe ER fluids here.

Electrorheological fluids are easy to make. They typically consist of micron sized particles suspended in a hydrophobic liquid such as mineral or corn oil. A simple suspension of corn starch[5] in corn oil works well. When an electric field is applied to such a mixture the initially randomly distributed particles form chains aligned with the field as in Figure 7.9. That is, the particles behave exactly as the systems we've already encountered in this section. From a

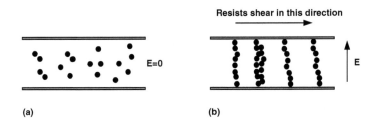

FIGURE 7.9: The behavior of particles in an electrorheological fluid. In (a) there is no applied field. In (b) the applied field causes the system to resist a shear in the direction shown.

technological standpoint, ER fluids are interesting because of the effect chain formation has on the behavior of the system under shear. If the system in Figure 7.9 (a) is sheared, it will behave like an ordinary fluid. If however, we attempt to apply a shear to the system in Figure 7.9 (b), the chains will now play an active role and resist our efforts to slide the top electrode. This type of control over a fluid can be useful in devices such as clutches and active shock absorbers.

From the standpoint of self-assembly there are several features of ER fluids worth noting. First, this is an example of a dynamic self-assembling system. When an electric field is applied, chains form, when the field is switched off, the chains collapse. The system is ordered only when dissipating energy. Next, it is again cooperation and competition between forces that creates order in the system. In ER fluids, electrostatic forces and hydrodynamic forces cooperate and compete as chains are assembled. Finally, recent studies of ER fluids have shown that when a field is applied the internal structure of an ER fluid can exhibit more order than simply the formation of chains. In [137, 138] Wen and Lu demonstrated that net-like structures, reminiscent of a cross-linked polymers, could be formed. They further showed that by varying the surface properties of the particles suspended in the fluid, the patterns formed could be controlled. The reader is referred to [137, 138] for further details.

7.3   Magnetically Driven Dynamic Systems

In the systems of the previous section, electrostatic forces, in competition or cooperation with gravitational and hydrodynamic forces, led to dynamically self-assembled structures. In this section, we focus on systems where *magnetic* and hydrodynamic forces cooperate and compete in the formation of dynamic self-assembled structures.

Here, we will describe aspects of the sequence of studies carried out by George M. Whitesides, Bartosz A. Grzybowski, and various collaborators and reported on in [55, 56, 57, 58]. The basic experimental setup used in [55, 56, 57] is shown in Figure 7.10. In these studies millimeter-sized magnetic disks

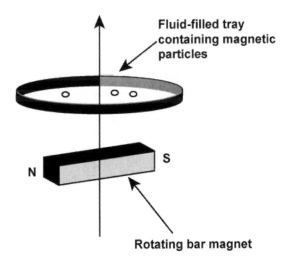

FIGURE 7.10:   The experimental setup for the magnetically driven dynamic self-assembling system.

were fabricated and placed in a circular fluid filled tray. The fluid was non-magnetic; a typical choice of fluid was a water and glycerine mixture. A permanent bar magnet was placed below the disk and aligned so that a line drawn through its center would pass through the center of the dish. The magnet was then rotated about this axis.

When the magnet was not moving, the particles in the fluid did feel a magnetic force. In this case, they would be attracted towards the poles of the magnet and would cluster in the fluid at a point above these poles. There they would remain forming largely disordered aggregates. Once the magnetic bar was set in motion the system came alive. Instead of being attracted to the poles as with the stationary magnet, the rotating magnetic field attracted the particles towards the axis of rotation. In addition, the magnetic particles became entrained to the rotation of the bar and began spinning about their centers. The rotation of the particles set the surrounding fluid in motion and created a repulsive hydrodynamic force between the particles.

The rotation of the magnet and its subsequent effects on the particles and the fluid in the tray endowed this system with the properties needed to dynamically assemble interesting structures. The magnetic forces were cooperative, drawing all of the particles towards the center of the tray. The hydrodynamic forces were competitive, pushing nearby particles apart. By changing the numbers of particles in the system the group was able to create a variety of dynamic structures. Sketches of these structures are shown in Figure

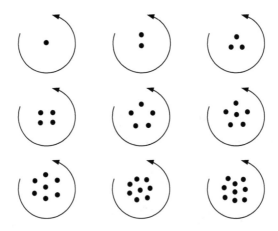

FIGURE 7.11: Dynamic patterns formed in the rotating magnetic disk system.

7.11. Note that when a single particle was used, it migrated to the center and rotated. When more than one particle was used, mutual repulsion between the particles balanced with the magnetic attraction towards the center. This forced the particles into a variety of geometric patterns. These patterns rotated as a whole about the center of the tray. The group also found that for certain values of the number of particles more than one dynamic structure was possible. When ten, twelve, or nineteen particles were used two different dynamic states were observed. For the cases of ten or twelve particles the system spontaneously switched between these states. When nineteen particles were used, the states were segregated by a threshold value of the rotational speed of the bar magnet. Finally, the group also showed that the spacing between the disks was only a function of the angular speed of the bar magnet. This indicates that the magnetic force on a particle was independent of the number of particles but that the hydrodynamic repulsive force increased with the particle's increasing angular speed.

In [58] the group extended the basic experimental setup of Figure 7.10 and showed that a modified version could be used to produce self-assembling micro-fluidic machines. This time, in addition to a rotating permanent magnet, the group embedded an array of electromagnets beneath the fluid layer in the tray. These electromagnets could be controlled by the user. They also modified the disk shaped particles of their previous experiments, this time creating small rotors. As before, when the permanent magnet was set into motion, the rotors were attracted towards the center of the tray, individually rotated, and repulsed one another via hydrodynamic forces. However, here once a dynamic structure had formed it could be further manipulated using the embedded electromagnets. The group fabricated three different functioning machines using this approach. In one of their machines seven rotors and nineteen electromagnets were used. By activating the electromagnets the group could freeze the rotors into the structure shown in Figure 7.12 (c). Note that each rotor continues to spin even when the overall structure is locked in place. This means that a flow is created in the fluid. It was this flow that the group sought to manipulate and use to build a working device. The action of their rotary "carousel" system is shown in Figure 7.12. Note the presence of a small circular container in these figures. When the carousel was locked in place with all of the rotors spinning, the container remained outside of the carousel. By selectively activating the embedded electromagnets, the group could break the array and cause the container to move to the interior of the carousel. This is shown in Figure 7.12 (b) and (c). The carousel then moved the container around the central axis till it reached the point shown in Figure 7.12 (e). Here, a small syringe filled the container with a dyed fluid. The carousel then moved the container up to the point shown in Figure 7.12 (f), finally ejecting it back into the bulk fluid. The carousel array was then back in its original configuration and ready to process additional containers.

The systems described in this section serve as another nice example of an engineered dynamic self-assembling system. Here, the role of both the attractive magnetic forces and the repulsive hydrodynamic forces is clear. In particular, these systems illustrate how system variables can be manipulated

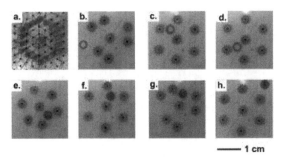

FIGURE 7.12: A self-assembling micro-fluidic machine. (a) shows the array of 19 electromagnets used in the system. Parts (b) through (h) show the system acting as a pump. The empty container is drawn into the array in (b), it is filled by (e) and ejected in (h). Reprinted with permission from Grzybowski et al., Applied Physics Letters, v. 84, 2004, pp. 1800. Copyright 2004 by the American Institute of Physics.

to create different dynamic patterns. It is interesting to compare the system described here with the system of Stambaugh et al. described in Chapter 5. According to our definition, the Stambaugh system, although static in its assembled state, is an example of dynamic self-assembly. Recall that the Stambaugh system used an externally applied magnetic field to confine a set of magnetic particles. When the magnetic field was switched off, the system collapsed from an ordered dynamic state to an ordered static state. This is similar to what we've seen above. Here, when the permanent magnet is not rotating, the system migrates towards a somewhat ordered static state. When the magnet is set in motion, the system switches to a dynamic state. The key difference between the system of Stambaugh et al. and the systems of this section is the addition of a second significant force to the system. In particular, the systems of this section made use of repulsive hydrodynamic forces. This added a layer of structure not present in the system of Stambaugh et al.

7.4 Mechanically Driven Dynamic Systems

Systems of particles suspended in a fluid can be vibrated electrostatically and magnetically. They can also simply be shaken. This mechanical driving of granular fluids brings us into contact with the vast area of research into the nature of granular media. Here, we briefly describe two simple studies closely related to the other systems considered in this chapter. The reader is directed to the Related Reading section at the end of this chapter for an introduction to the field of granular media and pointers to the enormous body of literature on this subject.

The first system we'll consider is due to Voth et al. [135]. The setup of the experimental apparatus is very similar to the apparatus used by the Aranson group pictured in Figure 7.8. The Voth et al. setup for mechanical shaking is shown in Figure 7.13. This group suspended tiny stainless steel spheres in a

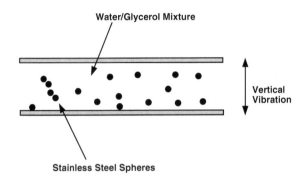

FIGURE 7.13:   The experimental apparatus for the vibrated granular system.

mixture of water and glycerol. The radius of the spheres was approximately 0.4mm. The mixture filled a short, fat, cylindrical aluminum tank. The tank was about 6cm in diameter and about 1.5cm in height. The top of the tank was sealed with a glass window so that high speed video could be captured from above. The entire apparatus was placed on an electromagnetic vibrator. Both the frequency and amplitude of the vibrations could be controlled.

As with previous systems, when there was no applied force, the particles simply distributed themselves randomly on the bottom plate. When the vibrator was turned on the particles hopped up and down, typically striking the bottom cell once each cycle. When a large number of particles was used, the group observed coarsening behavior like that in the electrostatic system examined earlier in this chapter. The initially randomly distributed particles clustered together and the clusters slowly coalesced into a large superstructure. To explain this behavior the group identified a hydrodynamic attractive force operating between the particles. Modelling the flow around a single particle, they showed that there was a steady inflow of fluid towards the equator of each particle. This inflow provided an attractive hydrodynamic force. The group also identified the presence of a hydrodynamic repulsive force in the system that acted only when large accelerations were applied by the electromagnetic vibrator. Using flow visualization techniques, they were able to observe recirculation zones near the particles. They speculated that the observed repulsive force was due to these recirculation zones, but could not demonstrate this conclusively.

The observation that the system could produce both attractive and repulsive forces led the group to study the behavior of systems consisting of only a handful of particles. When three particles were used, they anticipated that the particles would form a stable triangular structure in much the same way as the magnetic system of Section 7.3. However, the particles did not quite behave. When accelerations were low, a nearly triangular shape formed. But, when accelerations were increased, the system transitioned to a state with two particles clustered and the third particle dancing about in the distance. On occasion the third particle would wander back to the pair, ask for a turn to dance and form a new cluster while one of the previously paired particles was ejected into the distance. When seven particles were used the system again exhibited an acceleration dependent transitional behavior. For low accelerations, the group observed stable hexagonal structures. When accelerations increased, the structure again became time dependent with two central particles dancing inside a ring comprised of the five others. As with other systems described in this chapter, video is essential to fully appreciate the system's behavior. A link may be found in Appendix B.

The second mechanically driven dynamically self-assembling system we'll discuss is due to the group led by Harry L. Swinney. Details of this system may be found in [92]. In some sense, this group simply took a typical electrorheological fluid and shook it vertically. Remarkably, this led to fascinating behavior. The team worked with two different fluid mixtures. The first was a simple suspension of cornstarch in water. The second was a suspension of glass microspheres, of diameter between one and twenty microns, dispersed in water. In either case, the fluid was placed in a circular dish about 9.4cm in diameter. The cornstarch was poured to a depth of 0.5cm, the glass spheres were poured to a depth of 0.2cm. The top of the container was sealed with a clear glass plate and images were taken from the top. As above, the entire apparatus was placed on a mechanical shaker. The frequency and amplitude of the vibrations could be adjusted.

When the shaker was turned on the group observed a striking range of patterns on the surface. Most notable was the presence of persistent holes in the fluid. These holes, or cylindrical vacancies, penetrated entirely through the depth of the fluid and could last for the lifetime of the experiment. To better understand this phenomenon, the group mapped out a phase diagram for the system. Here, the different behaviors are a function of the forcing frequency and acceleration. When the acceleration was low, holes were unstable. They found that they could initiate hole growth by shooting a puff of air towards the surface of the fluid but that these holes would rapidly close. For larger accelerations, and the right choice of frequency, the group identified regions of the phase diagram where holes would form, adjust to a well-defined size, and persist. In still other regions of the phase diagram, the group observed what they called "delocalized" behavior. In this region, holes would form a hump on their upper rim. This hump would grow out of the fluid like a finger, eventually reach a maximum height, and then topple back into the fluid. The toppling excited other regions of the fluid and eventually the entire surface was filled with evolving voids and vertical structures. Again, video is essential to truly appreciate this genuinely weird behavior. A link may be found in Appendix B.

7.5 Self-Propelled Systems

In the systems we've considered thus far, the driving force has come from an *externally* applied field. In the system described in this section, it is the interaction of structured particles with the environment that provides the driving force. This system is yet another creation of the Whitesides group [65] and was inspired by complex biological systems such as swarming bacteria and schooling fish. Their intent was to capture the complex behavior exhibited by large collections of independent agents in an engineered self-assembling system.

The group began by fabricating PDMS tiles as in their previous studies. Once again, the hydrophobic and hydrophilic properties of the tiles could be tailored. This time, an additional modification was made to the tiles. Each tile was outfitted with a small platinum coated glass "fin." This fin was oriented perpendicularly to the face of the tile and attached to the tile by a steel pin. A sketch of the tiles appears in Figure 7.14. The tiles were placed on the surface of a liquid. Here, the liquid was a mixture of water and hydrogen peroxide. The platinum fin was immersed in the liquid while the upper face of the tile remained in the air. When the fin made contact with the hydrogen peroxide solution, the platinum coating catalyzed the decomposition of the hydrogen peroxide into water and oxygen. This caused small gas bubbles to form at the fin surface. The ejection of these bubbles from the fin provided a locomotive force for the particle. The group found that their fin equipped particles could continue to move with almost constant velocity for several hours.

Note that the shape of the particles in Figure 7.14 lacks symmetry. This implies that when propelled through the hydrogen peroxide solution the particles will rotate. In their experiments, the Whitesides group used particles shaped like those in Figure 7.14 and particles with the same shape, but flipped upside down. This meant that their system contained particles that swam both clockwise and counterclockwise. The wetting pattern around the edges of the particles was designed so that particles could bind pairwise. Working with a single pair of particles, chosen to swim in opposite directions, they found that stable binding did occur. The pair formed a two particle complex that remained connected and rotated. When a collection of ten particles was

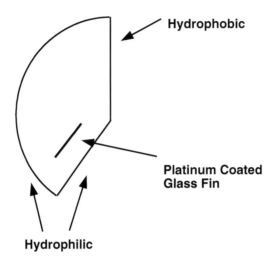

FIGURE 7.14: Design of self-propelled particles. This is a view from the bottom. The platinum coated fin is immersed in the liquid layer.

used, the particles still formed stable particle pairs, but not with one hundred percent yield.

The system described here constitutes a proof of concept. The Whitesides group was able to demonstrate that it was possible to design a dynamic self-assembling system without using an externally applied field. Unfortunately, they have not yet been able to carry out experiments using large numbers of their self-propelling particles. They speculate that such experiments will lead to more complex emergent behavior. Simulations of such systems support this speculation. For one such study the reader is directed to [29].

7.6 Smart Particles

The systems of this chapter as well as those of Chapters 5 and 6 are quite impressive. But, we must face one fact. All of the particles used in all of our clever engineered self-assembling systems are *dumb*. We may give them fancy names like "structured particles," we may paint them with hydrophobic and hydrophilic stripes, we may give them pretty shapes and decorate them with magnets. But, they're still dumb.

In sharp contrast, nature's particles are *smart*. When we discussed proteins in Chapter 3, we described how a protein folds and adopts a particular conformation. Proteins are also able to change shape, that is, change this conformation, in response to interactions with other proteins or the environment. We saw this behavior in the tobacco mosaic virus when self-assembled protein washers switched to a lock-washer geometry in response to interaction with RNA. We call this behavior *conformational switching*. This change in shape allows nature's particles to change how they bind and how they interact with the environment. Nature's particles are smart.

Now, to be fair, we have seen some particles that can undergo conformational changes. For example, in the electrostatic systems of this chapter, particles were able to change their state by changing their charge. But, they don't do so intelligently. Yes, they do change in response to other particles and they do change in response to the environment, but they are mere passive participants in this process. In contrast to particles like the TMV disk proteins, that only change conformation when the right strand of RNA appears, these particles still appear dumb.

So, how do we move towards engineered systems that use smart particles? One approach to this has been pioneered by Eric Klavins [14, 73] and will be discussed in this section. However, before we examine the Klavins system, let's revisit some of our earlier self-assembling systems and see what features smart particles might have and how they might help us design better systems.

First, let's consider how smart particles might help us build better systems. Imagine we returned to one of the tile based systems of Chapter 6 and constructed a set of triangular tiles like those shown in Figure 7.15. We'll assume

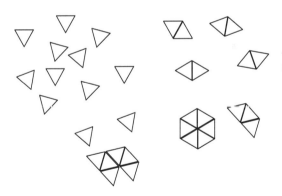

FIGURE 7.15: A collection of self-assembling triangles forming different shapes.

that all of the edges are capable of binding. We might accomplish this magnetically or by using the wetting properties of our tiles. If we placed our triangles in the proper environment, we'd see them start to form triangle complexes like those shown in the figure. We'd see triangle doublets appear, triplets, chains, and perhaps even hexagonal structures. But, suppose our goal was to form only hexagons? Well, in this system, we'd be faced with a serious yield problem. We might obtain a few hexagons, but our experience with similar systems in Chapter 6 leads us to believe that our yield would be low indeed. Now, we have developed some tools to help us attack this problem. We could turn to the use of templates, we could change the patterns of binding forces on the particle edges, or we could tinker with our particle shapes. Each of these approaches would yield some success. But, now suppose our goal was more ambitious. Suppose we wanted to force all of the triangles to form hexagons, remain in this state for some period of time, and then return to a state of unbound triangles. The techniques we've see thus far are no longer of much help. We could attempt to use some of the ideas outlined in this chapter to build such a system, but wouldn't it be much simpler if somehow our particles could decide what state they were in and make the switch from hexagons to triangles on their own? Wouldn't it be nice if our particles were *smart*? If the triangular particles of Figure 7.15 were smart, they could perhaps sense what sort of structure they were bound to and decide on their own to remain bound or not. Endowing our particles with such an ability would clearly make the design of dynamic self-assembling systems much easier.

Let's consider this concept of smart particles in yet another context. Recall that in Chapter 2 we discussed the process of polymerization. Let's imagine a highly simplified polymerization process that consists of polymers of only three lengths. We can imagine polymer chains consisting of one, two, or precisely three monomers as being the elements of our system. We'll pretend that no catalyst is needed for our system and simply assume that two simple reactions govern our polymerization process. Namely,

$$P_1 + P_1 \longrightarrow P_2 \tag{7.2}$$
$$P_1 + P_2 \longrightarrow P_3.$$

Using the Law of Mass Action we can write down equations governing the concentrations of our reactants. Assuming a reaction rate of k for all reaction we find that the concentrations, denoted by $p_i(t)$, satisfy

$$\frac{dp_1}{dt} = -kp_1^2 - kp_1p_2 \tag{7.3}$$
$$\frac{dp_2}{dt} = kp_1^2 - kp_1p_2$$
$$\frac{dp_3}{dt} = kp_1p_2.$$

Even though this system is nonlinear, it is not hard to see how the p_i evolve. Clearly, p_1 only decreases and p_3 only increases. The behavior of p_2 depends on initial conditions, but, even if p_2 increases initially it will eventually tend toward zero. This is not surprising. Our reactions say that the p_1's combine to form p_2's, the p_2's combine with p_1's to form p_3's, and nothing else can happen.

Profile - Eric Klavins

Some people use air hockey tables to play air hockey, Eric Klavins uses his as a platform for his army of programmable self-assembling robots. Klavins, an electrical engineer at the University of Washington, is a member of the new generation of self-assembly pioneers, taking a classical education in computer science and engineering and applying it to the difficult and challenging task of designing and building nontrivial self-assembling systems. In 2004, Eric received a prestigious National Science Foundation Career Award based on his work in programmed robotic self-assembly. Since then, Eric and his graduate student, Nils Napp, have built triangular, programmable robotic "parts" that live on an air hockey table, randomly mixing, and self-assemble into desired structures. The parts bind magnetically and are endowed with the ability to communicate with one another and make decisions concerning binding based on a locally stored graph grammar. Eric has been a leader in using graph grammars as a tool to model and control self-assembly. Photos of the Klavins' robots are shown below while fascinating videos of the assembly process are available on his web page. The link may be found in Appendix B.

In a recent conversation, Eric commented on the blind-spots holding back development in the field of self-assembly. He writes:

> *I think we keep confronting massive combinatorial state spaces that lack convenient models. Low level models are too complex, while continuum models miss the details. People avoid systems that generate such state spaces and focus too much on simple self-assembly, essentially crystallization. We need to focus on the fundamentally new algorithmic possibilities of self-assembly.*

Currently, Eric and his group are extending their approach to MEMS and DNA self-assembly. They are working to apply their ideas about active, programmable assembly, developed on their air hockey table, to micro and nanoscale systems.

FIGURE 7.16: Eric Klavins' self-assembling robotic parts. The inset shows a self-assembled structure. Credit: Eric Klavins and Nils Napp.

Eventually, most of what we form must be polymers of length three. But suppose that what we really wanted was to form polymers of length two. Then, we'd again be faced with a yield problem. We saw in Chapter 2, that the distribution of polymer lengths evolved with time. The same is true here. We could analyze our system in more detail and then stop the polymerization process at the instant when the yield of the p_2's was maximized. But, if our particles were smart, we could increase our yield of p_2's and avoid worrying about precisely when to stop the process.

To see how this might work, imagine that each of our monomers was able to sense its state. That is, suppose periodically, each monomer checked whether it was still a monomer, a dimer, or a trimer. Further, assume that each particle could make a decision about its binding based on the state identified. If a particle found it was a monomer or a dimer, we'd want it to do nothing. If however, it found it was part of a trimer, we'd like it to unbind and destroy the trimer. However, there is a complication. If all of the subunits simultaneously decide to unbind, we'd produce monomers when we could be producing the desired dimers. So, in addition to being able to identify their state, we want our particles to communicate and somehow mutually decide what to do. For the polymerization system we wish to add a new reaction pathway that Klavins calls a *programmed reaction pathway*. For our system this is

$$P_3 \longrightarrow P_1 + P_2. \tag{7.4}$$

If we included this reaction and assumed it occurred at some programmed rate, k_p, our system of differential equations would become

$$\frac{dp_1}{dt} = -kp_1^2 - kp_1p_2 + k_pp_3 \tag{7.5}$$

$$\frac{dp_2}{dt} = kp_1^2 - kp_1p_2 + k_pp_3$$

$$\frac{dp_3}{dt} = kp_1p_2 - k_pp_3.$$

Now the behavior of our system is closer to what we desired. The concentrations of p_1 and p_3 no longer evolve monotonically. Rather, the breakup of the p_3's ensures that new p_1's are produced eventually leading to the desired p_2's. Note that this system can be optimized by choosing k_p. We invite the reader to investigate this further in the exercises for this section.

The protocol outlined above, that is, the use of particles that can both sense their state and collectively make decisions about binding, was implemented by Eric Klavins et al. in [14, 73]. Klavins et al. designed triangular programmable particles and allowed them to mix and interact on an air-hockey table. A photograph of one of the team's particles appears in Color Plate 11.11. Each triangular particle contained three controllable magnetic latches, three infrared transceivers, and a logic circuit. The latches and transceivers were arranged symmetrically so that the triangular parts could bind as in our thought experiment of Figure 7.15. Binding was accomplished via the use of permanent magnets. Each magnetic latch on the particle actually consisted of three permanent magnets. One of these, the central magnet, was fixed in place with its north face protruding outwards from the particle. Surrounding this were two movable magnets whose position could be adjusted by a motor. In the default state, the two movable magnets point with their south faces facing outwards. When two particles come sufficiently close, they bind. The fixed magnet of one particle will attach itself to one of the movable magnets of the other particle. However, particles could unbind. If a pair of particles decided to unbind, they would each use an onboard motor to rotate their movable magnets by 180 degrees. This would force the particles apart. The moveable magnets would then return to the default state, allowing each particle to again bind freely.

In addition to the mechanical ability to bind and unbind, the particles also needed to be able to sense their state, communicate with their neighbors, and collectively make decisions about binding. To implement this aspect of the smart particle design, each particle was equipped with a logic circuit and an infrared transceiver. The transceiver allowed particles to communicate. The logic circuit allowed the particles to decide whether or not to remain bound. The circuit made decisions based on a *graph grammar*. Graph grammars represent a powerful approach towards understanding and modelling self-assembly. We'll revisit graph grammars in Chapter 9. Here, we simply note that each particle defined its state in terms of the position of its three latches. This state and the states of nearby neighbors were then examined by the logic circuit. The logic circuit essentially translated these states into the language of a predefined graph grammar, examined the graph grammar to determine how to act, and communicated this information back to the latches. Notice that the particles were programmable. The graph grammar by which the logic circuits made decisions was stored in each particle's internal memory. But, this graph grammar could be changed and hence this system really achieved *programmable* self-assembly. In [73], Klavins et al. showed how to specify different graph grammars to achieve different assembly goals. In fact, they even showed how to specify a graph grammar so that their robotic triangles would self-assemble into the hexagonal structures we originally considered at the start of this section.

7.7 Chapter Highlights

- Dynamic self-assembling systems are those that produce ordered structures that remain ordered only so long as the system is dissipating energy.

- For dynamic self-assembly to occur their often must be a competition between particle interaction forces. In our prototype system, particles are electrically charged and can hence attract or repel. Charged particles can change their charge and hence change how they interact with other particles.

- The competitive and cooperative efforts of electrostatic, hydrodynamic, and gravitational forces in a particle-fluid system can lead to the formation of one, two, and three dimensional spatial and temporal structures.

- Magnetic forces in competition with hydrodynamic forces may also be used to create dynamic self-assembling systems. This competition can be used to create structures consisting of only a handful of particles, or massive structures capable of changing the bulk behavior of a fluid.

- Dynamic self-assembling systems may also be formed using purely mechanical driving. Here, it is a complex interaction of attractive and repulsive hydrodynamic forces that gives rise to structure.

- The use of an externally applied driving force may be avoided by creating particles that are *self-propelling*. These particles interact with their environment to produce a motive force. Experiments with self-propelling particles may yield insight into collective biological behavior such as swarming and schooling.

- To truly approach the abilities of nature we must turn to *smart particles*. Smart particles mimic proteins and are able to undergo *conformational changes* in response to other particles or the state of the environment.

7.8   Exercises

Section 7.2

1. Consider a single conducting particle in a dielectric fluid placed between two electrodes. Assume the electrodes are parallel infinite plates. Assume the effect of the particle on the field is negligible and compute the electric field in the gap. Now, use this to write down an equation of motion for the particle. Include drag forces on the particle in your model. You may assume the particle is spherical so that Stokes' Law applies. Under what conditions will the inertial terms be negligible?

2. Repeat the problem above but this time include the effect of gravity. Assume that the gravitational force points in a direction perpendicular to the electrodes. How much charge must a particle acquire if it is to be able to move upwards against the force of gravity?

3. One approach to simulating the systems discussed in this chapter is to take an agent based approach. In this approach, each particle is treated as an autonomous agent interacting with its neighbors via simple rules. Build a one-dimensional simulation of particles moving between two electrodes using such an approach. Assume each particle is either in state +1 or −1, that when particles collide with particles in the opposite state they switch states and set their velocities to zero and that particles accelerate towards the electrode with the opposite state. One electrode is to be permanently in state +1, the other in state −1.

Section 7.3

4. The magnetic self-assembling systems of Chapter 5 produced structures resembling those of this chapter. In this chapter we've briefly compared these systems. Return to Chapter 5 and carefully compare the magnetic systems described there with the magnetic systems described here. Explain exactly how these systems differ and exactly how they are alike.

5. Contrast the combination of attractive and repulsive forces used by the systems in this section with the Hosokawa et al. system of Chapter 6. Why does one combination lead to static self-assembly and the other to dynamic self-assembly?

6. In the exercises in Chapter 2, you made use of MIT's Star Logo system to simulate Diffusion Limited Aggregation. Star Logo is equally useful for exploring the behavior of the systems described in this section. Design a Star Logo simulation where the "turtles" are attracted to a common point in space, but repulsed from one another. Does the outcome of your simulation resemble the structures formed by the systems in this section?

Section 7.4

7. Modify your agent based simulation of electrostatic self-assembly to construct a model of a mechanically driven system. Again, suppose your system is one-dimensional. How should the states of your particles be specified? How does the motion of the particles in this simulation differ from your last simulation?

8. For systems in other sections of this chapter we identified the particles, the binding force, the environment, and the driving force in the system. Do the same for the systems of this section. Further, discuss the forward, backward, and yield problems for these systems.

Section 7.5

9. Again, Star Logo is a wonderful tool for creating simple models of systems of autonomous agents. Using Star Logo, create a simulation of the

system described in this section. Be sure to include the biased rotational motion of the particles constructed by the Whitesides group.

Section 7.6

10. Consider the two different systems of ordinary differential equations governing the polymerization process of this section. Either numerically, or analytically, analyze these equations. Compare the behavior of solutions for the two different systems. Discuss how k_p can be chosen to optimize the yield of p_2.

7.9   Related Reading

The text by Thomas B. Jones is the best reference on the behavior of particles in an electric field.

T.B. Jones, *Electromechanics of Particles*, Cambridge University Press, 1995.

The text by Larson, also mentioned in Chapter 2, provides a nice introduction to electrorheological and magnetorheological fluids.

R.G. Larson, *The Structure and Rheology of Complex Fluids*, Oxford University Press, 1999.

The edited collection by Halsey and Metha gives a good introduction to granular flow.

T. Halsey and A. Metha, *Challenges in Granular Physics*, World Scientific, 2000.

If you are interested in the behavior of mechanically vibrated granular media, the place to start is with the work of the physicist Harry L. Swinney. A link to his web page, which contains most of his publications, may be found in Appendix B.

In some cases the hydrodynamic forces between particles in a fluid can be computed. The classic results in this area may be found in the book by Batchelor.

G.K. Batchelor, *An Introduction to Fluid Dynamics*, Cambridge: Cambridge University Press, 1967.

Many computational simulations of systems like the system of self-propelled particles described in this chapter have been carried out. Such research forms a large part of the field of artificial life. A good introduction to this field is the book by Adami.

C. Adami, *Introduction to Artificial Life*, Springer, 1998.

7.10 Notes

1. I'm assuming your head is not bald. If it is, you'll need to conjure up a friend for this thought experiment.

2. The system described here was fabricated in the MEC Lab at the University of Delaware under the direction of the author. Several students, D. Cargill, T. Fleetman, and O. Breslauer, participated in this work.

3. We used ACME brand corn oil as the dielectric fluid.

4. This is the essence of the classic *Franklin's Bells* experiment due to Benjamin Franklin.

5. Corn starch is fascinating stuff. A suspension of corn starch in water produces a shear thickening fluid. If you fill a pool with this mixture you can actually run across the surface without sinking.

Chapter 8

DNA Self-Assembly

It is a strange model and embodies several unusual features. However, since DNA is an unusual substance, we are not hesitant in being bold.

James D. Watson

8.1 Introduction

The quote above is taken from a letter to a friend written by the co-discoverer of the structure of DNA, James D. Watson, a month before their discovery was made public.[1] Watson got it right. DNA *is* strange, it *is* unusual, and harnessing its power has required and will require truly bold acts by scientific thinkers in every discipline. Yet, that's where the fun is, and that's where the promise of self-assembly truly comes alive.

In this chapter we examine DNA based self-assembly. We'll look at the progress that's been made, highlight the pitfalls and problems, and see some of the tremendous opportunity for nanoscale engineering that is made possible by DNA. We begin in Section 8.2 with a brief review of DNA's structural and chemical properties. We'll review the important concept of *base pairing*, sometimes called Watson-Crick base pairing, that is responsible for DNA's ability to self-replicate and its usefulness as a self-assembling structural material. In Section 8.3, we'll examine some of the early successes in using DNA as a self-assembling construction material. We'll learn about *sticky ends* and *branched junctions*, two forms of DNA that make construction possible. We'll see how by using sticky ends and branched junctions various groups have succeeded in self-assembling three dimensional nanoscale polyhedra from DNA. We'll also see some of the problems they encountered along the way, and learn how many of these obstacles are being overcome. We'll see how the common problem of rigidity is overcome through the use of the DNA *double crossover* molecule (DX). The DX molecule will play a central role in Section 8.4 where we examine *DNA tiles*. We'll see how these tile systems are similar to many of the systems of Chapter 6 and we'll see why DNA tiles succeed where macroscale tiles often fail. This section and Section 8.5 will also provide us with examples of *programmable self-assembly*. We'll see how changing the sequence of base pairs on sticky ends, or changing a family of tile types amounts to programmable control over self-assembled structures. We'll also see how structures formed from tiles can be used as templates for functional nanodevices. In Section 8.6 we'll see how the promise of DNA tiling has been vastly extended through the method known as *DNA Origami*. In this technique, arbitrary two dimensional shapes can be self-assembled from a long single strand of DNA. In turn, these complex shapes can be used as tiles in self-assembled DNA tile structures. In Section 8.7, we'll see how DNA can be used directly as a template for the assembly of nanostructures. We'll examine a DNA template design for a nanoscale transistor, a key component of digital electronics, and one that has already been built using DNA based self-assembly. Finally in Section 8.8, we'll examine DNA based self-assembly in the context of what we've learned in the previous seven chapters. While DNA is strange, and it is unusual, we'll see that DNA based self-assembly presents us with the same obstacles and challenges we've encountered before.

8.2   DNA - Nature's Ultimate Building Block

You can't get away from DNA; it is truly nature's molecular pop-star. In the fifty or so years since Watson and Crick illuminated the structure of nature's instruction manual, DNA has come to pervade popular culture. Images such as Color Plate 11.8 grace the cover of magazines, books, and even compact discs. Countless companies embed the DNA double helix in their corporate logo. The terms "DNA fingerprinting," "gene sequencing," and "DNA testing," have entered the popular lexicon. In 2003, a Harris poll even showed amazingly, that more than sixty percent of American adults could correctly answer the question "What is DNA?"

So, thus far in this book, when we've mentioned DNA, we've assumed that you have some working knowledge of the DNA molecule. But, before we can go further and discuss how DNA is used in self-assembly, we need to review the structure of DNA in a bit more detail.

DNA is an acronym for *deoxyribonucleic acid*. The term *deoxyribose* describes the cyclic sugar molecule that makes up DNA's backbone. The structure of deoxyribose is shown in Figure 8.1. The term *nucleic* describes the fact that DNA is found in the nucleus of the cell. Hence the term deoxyribonucleic. The sugar molecules in DNA are linked via phosphoric acid units, hence the term *acid*. So, DNA is a long-chain molecule, a polymer, whose backbone is built from sugar molecules linked together by acid units. But, each sugar unit in DNA is also linked to one of four heterocyclic bases, adenine, guanine, cytosine, or thymine. The structure of each of these bases is shown in Figure 8.2. It is, of course, these bases, or *nucleotides*, usually de-

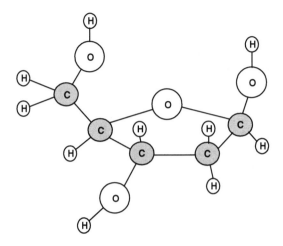

FIGURE 8.1: The structure of deoxyribose.

noted simply A,G,C, and T, that encode the genetic information carried by DNA. The basic structural unit of DNA is shown in Figure 8.3.

Now, DNA does not naturally exist as a single strand polymer. Rather, the basic structural unit of Figure 8.3 forms a long repeating chain with variations in the base unit and then binds to a complementary strand. It is this double strand that twists and forms the familiar double helix, Color Plate 11.8. The complementary strand is determined by *base pairing*. Each of the four bases, A,G,C, and T, bind selectively to a complementary base. In particular, A binds to T and G binds to C. So, given the sequence along one backbone, say *AATGC*, its complement, *TTACG*, is automatically determined.

This base pair structure is at the heart of DNA's ability to self-replicate and DNA's ability to carry information. Self-replication is possible because of the selective recognition of base pairs. If we begin with a single strand of DNA, a complementary strand can be built along this backbone. If the two strands are pulled apart, the complementary strand can then be used as a template to construct a copy of the original strand.

The information DNA carries is in the form of instructions for building proteins. Recall that proteins are built from amino acids and that living systems use approximately twenty different acids in building proteins. To encode for these twenty different amino acids, DNA uses triplets of the bases, A,G,C, and T. That is, each amino acid is identified by a group of three bases. There are 64 possible such groups, and hence sufficiently many to encode for all of the amino acids. Not all triplets encode for an amino acid. Some triplets instead serve as control instructions. For example, a *stop codon*, tells the cellular machinery when it has reached the end of a protein and can cease construction.

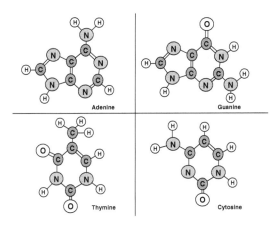

FIGURE 8.2:　The chemical structure of the base units of DNA.

8.2.1　Sticky Ends and Branches

In its naturally occurring double helical structure, DNA is not very useful as a building material. One essentially has long, not terribly rigid, sticks. A good image to hold in your mind is two strands of cooked spaghetti wound together in a helix. Further, at first glance, we have no way to attach strands of DNA and build larger structures. But, DNA can be pushed further. By using *sticky ends* and *branched junctions*, DNA can be turned into a useful nanoscale building material.

Sticky ends occur when one strand of a DNA double helix juts out past the end of the other. Again, imagine your strands of spaghetti where this time one strand is longer than the other. An example is shown in Figure 8.4. Here, on the left, we see the right end of a double strand DNA molecule with the lower strand continuing on past the end of the upper strand. This short protruding strand is the *sticky end*. This sticky end is available to selectively bind to a variety of molecular structures. Again, consider Figure 8.4. To the right we see two double strand DNA molecules, each with a sticky end. In this case, the top strand of each continues past the lower strand. The upper double strand DNA molecule has a sticky end whose bases form the complement for the bases of the sticky end of the DNA strand on the left. Hence, these two can recognize one another and bind. The lower right hand DNA molecule also has a sticky end, but the sequence of base pairs does not match up with those of the molecule on the left. Hence, this piece of DNA cannot bind with the DNA on the left. This notion of sticky ends allows researchers to insert strands of DNA into precise locations in circular strands of DNA known as *plasmids*. This is the basis for the field of genetic engineering.

But, here we're interested in DNA as a construction material. The notion of a sticky end offers some hope, we can now take our DNA sticks and bind them

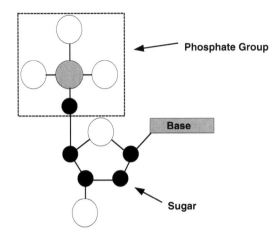

FIGURE 8.3: The basic structural unit of DNA. This unit repeats in a chain.

end to end to make a really big stick, or insert strands of DNA into a circular DNA loop to make a larger loop, but from the point of view of construction, not much else. We need the notion of *branched junctions* to truly make DNA construction possible.

In the cell, DNA does not always remain wrapped up in its double helical structure. If it did, it would not be of much use. Periodically, DNA must unwind and uncouple, in order for replication to occur and for genetic instructions to be delivered. When DNA unwinds it can form a *branched structure* such as the one shown in Figure 8.5. If two of these branched structures, with the right complementary sequences, come together, DNA can form a *branched junction*.

A typical branched junction is shown in Figure 8.6. Note that the location of the branch point need not remain fixed. The sequence of the upper left hand strand in Figure 8.6 matches that of the lower right hand strand. Similarly, the sequence of the upper right hand strand matches the lower left hand strand. Further note that the upper left hand strand is the complement of the upper right hand strand *and* the lower left hand strand. Again, similarly, the lower right hand strand is the complement of both the upper right hand strand and the lower left hand strand. Because of this symmetry, the branch point can slide around.

To get a feel for this, imagine a simple analogous situation. Suppose we had four strips of velcro. Let's assume two of those strips consist of "hooks" and the other two "loops." Suppose we arranged our four velcro strips like the four strips of a DNA branched junction. The result would resemble Figure 8.7. Notice that the hooked strands are located in the upper left and lower right and the loop strands are located in the upper right and lower left. Again,

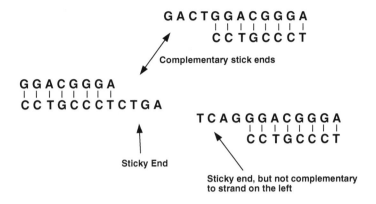

FIGURE 8.4:   DNA with sticky ends. All three strands have sticky ends, but only the left strand and upper right strand will bind.

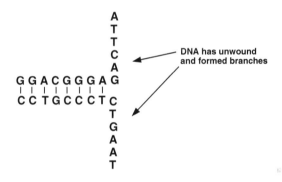

FIGURE 8.5:   The branched form of DNA.

this is just the same as the arrangement of our DNA strands in Figure 8.6. Clearly, we could slide this velcro junction and relocate it wherever we please. In this situation, the hook strands don't care where along the loop strands they bind to, just so long as there are loops.

To make junctions that don't move we need to break the symmetry of the structure of Figure 8.6. Fortunately, DNA is not like velcro. DNA's hooks *can* be made to care to which loops they bind. This is precisely the role of DNA's base pairs. Figure 8.8 shows a stable branched junction. Note that this time, the upper left and lower right strands are not the same. Neither are the upper right and lower left strands. Rather this time, the code along the strands has been chosen in a very particular way. We can imagine starting with two DNA double helices. We choose the code along one helix so that when it is unwound to a point it complements the code on the other helix exactly. But, beyond this point, we no longer allow the strands to be complementary. Instead, we

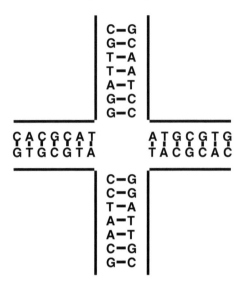

FIGURE 8.6: A mobile DNA branched junction.

vary the sequence of both helices so that beyond this point they no longer match. Note that this idea allows us to place a branched junction at any point along the length of a pair of DNA helices. We simply match to the desired point, and then cease matching beyond that point.

8.3   Cubes and other Polyhedra

In 1991, Junghuei Chen and Nadrian C. Seeman demonstrated the feasibility of using branched junctions to build nanoscale DNA structures with their fabrication of a DNA cube [26]. Since that time, Seeman's group as well as other groups worldwide have shown how to extend that idea to the fabrication of a truncated octahedron, a regular octahedron, Borromean rings, and even DNA knots [147, 122, 90, 117].

In their construction of a cube, Chen and Seeman actually made use of junctions that differ from the one in Figure 8.8. If you think about a cube for a moment, you'll see why. At the corners or vertices of a cube, three edges come together, not four. If we attempted to work with junctions like the one shown in Figure 8.8, we'd always have one extra edge to deal with. So, instead, Chen and Seeman engineered their DNA strands so that at the corners they formed the three armed junction of Figure 8.9. The edges of their cube consisted of twelve equal length double helices. The edges were short, each one allowed for

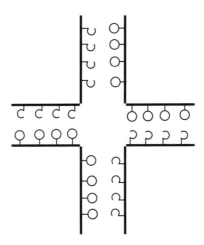

FIGURE 8.7:   The velcro hooks and loops analogy for mobile branched junctions.

only two turns in the DNA double helix. This meant that each edge was rigid. Chen and Seeman designed their DNA sequences such that the cube structure would be self-assembling. Here, this still meant that a number of intermediate steps were necessary. Essentially, Chen and Seeman self-assembled the faces of the cube and then used the process of ligation to connect the faces together. When a face self-assembled, it contained protruding sticky ends that allowed Chen and Seeman to make these face to face connections. At the end, they were left with the first nanoscale polyhedra, constructed entirely from DNA.

Chen and Seeman's cube did however suffer from one significant defect. It was floppy. If you've ever tried to build a cube, say using soda straws and balls of clay, you'll understand why. A cube is not mechanically rigid. It flexes, it flops, and it falls over. To remedy this, two approaches are possible. One approach would be to simply build something else. Other polyhedra are mechanically rigid. As we'll see in a moment, many groups have built other polyhedra, for this and other reasons. The other approach is to strengthen the joints. Recall that in their cube, they used very short strands of DNA to build the edges, hence these edges were very stiff. The difficulty lay in the joints. This is the exactly the same problem of mechanical rigidity that you'd encounter with a soda straw and clay construction. The straws won't buckle, rather the joints will flex.

To make rigid junctions with DNA, Seeman's group made use of a DNA double-crossover molecule (DX). In this structure, a pair of DNA molecules are aligned side by side, but there are strands crossing between the pair that tightly link them together. The basic structure is shown in Figure 8.10.

To this pair one can add a junction and obtain the so-called DX+J structure.

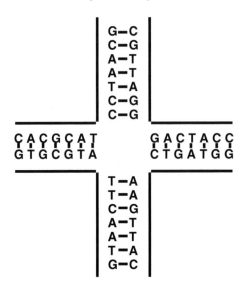

FIGURE 8.8: A stable branched junction.

Using these structures, the Seeman group has been able to build other three dimensional nanoscale objects such as a truncated octahedron that are indeed rigid. The reader is directed to [147] for further details on these constructions.

Chen and Seeman's cube suffered from one additional problem and until recently it was a problem shared by all DNA based nanoscale polyhedra. Namely, the Chen and Seeman cube and all other polyhedra were difficult to replicate. Ordinarily, DNA lends itself readily to self-replication. This is after all, one of its major functions. This ability to induce DNA to self-replicate is at the heart of the *polymerase chain reaction* (PCR) upon which much of modern biotechnology is based. Yet, the structure of branched DNA is different than that of ordinary DNA, and one consequence of this is that it does not easily self-replicate. Further, recall that the Chen and Seeman cube was built in steps and at each of these steps, faces of the cube had to be tied together. Even if a face could be easily replicated, ligation would still be necessary.

This difficulty was overcome in 2004 by the group led by William M. Shih. To accomplish this, Shih's group showed how to construct a self-assembling DNA polyhedra using a *single strand* of DNA assisted by short helper strands. These helper strands are complementary to short regions of the main strand, and in a cross-over motif add structural rigidity to the assembled polyhedra. Their main strand, 1700 base pairs long, was readily amenable to reproduction using the standard tools of molecular biology. In particular, standard PCR methods could be used to make arbitrarily many copies of their strand quickly and easily. And yet, in a very simple denaturation-renaturation procedure the

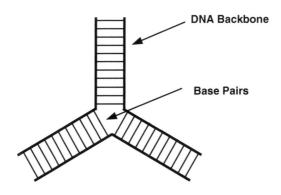

DNA Backbone

Base Pairs

FIGURE 8.9: The basic branched junction used by Chen and Seeman in building their DNA cube.

same strand would self-assemble into a three dimensional octahedra. Details of the Shih system may be found in [122].

One more problem, that until recently, had plagued the construction of self-assembled DNA polyhedra was the problem of yield. When Chen and Seeman self-assembled their DNA cube, the process involved three stages, several intermediate purification steps, and at the end, produced a yield of only one percent. Even later more refined, constructions by the Seeman group, such as the truncated octahedron [147] still had a disappointingly low yield of around one percent. The problem was one of local energy minima. Even though DNA binding is highly specific, the length of the strands and the variety of the bases virtually guarantees that there will be more than one stable way to put the basic pieces together. That is, even when the target structure is a global energy minimum, there are other nearby structures with only slightly higher energies. Every time a collection of pieces gets trapped in a nearby local minimum, your yield decreases. In 2005, Goodman et al. [51] demonstrated the construction of a family of DNA tetrahedra in a one-step process with a yield of almost ninety-five percent. To accomplish this feat, Goodman et al. used four short single strands of DNA. Their assembly process was simple, the strands were mixed in solution at 95°C, the solution was cooled to 4°C in about thirty seconds, and the product examined. The group designed their four DNA strands to interact in a hierarchical fashion. As the temperature of the solution dropped, strands would bind pairwise, as the temperature fell further, these pairs of strands would form the tetrahedra. They speculated that this hierarchy was responsible for the high yields they observed. Color Plate 11.13, shows atomic force microscope images of their assembled structures, as well as a schematic of the assembly sequence.

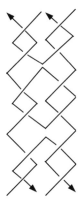

FIGURE 8.10: The basic structure of a DNA double-crossover molecule. Two complete DNA helices lay side by side and are joined as strands from one helix cross over to the other helix.

8.4   DNA Tiles

In 1998, Erik Winfree, Furong Liu, Lisa A. Wenzler, and Nadrian C. Seeman realized that the DX molecules introduced above could be used to design and fabricate *DNA tiles* [141]. Recall that in Chapter 6, we discussed several ways to self-assemble artificial crystals using specially designed tiles. We also noted that there is a connection between tiling and computation and that this connection has the deeper implication of connecting computation and self-assembly. In Chapter 6, we saw one attempt to exploit this connection when we examined Paul Rothemund's efforts to compute using capillary forces. The intent of Winfree et al. was identical. They sought to design DX molecules that would exhibit the selective binding necessary to achieve computation through self-assembly.

They began by designing the simplest possible nontrivial tile set. Their tile set consisted of just two different tile types. Each tile was a DX molecule, used four strands of DNA, and left sticky ends at both the right and left ends of the tile. The tiles were short, one being 36 base pairs in the length, the other 47 base pairs in length; this ensured that their tiles were rigid. Abstractly, we can picture their two tile types as in Figure 8.11. Notice that there are four distinct shaded regions on each tile. These represent the sticky ends. The sequence of bases on the sticky ends is chosen so that they will only bind to complementary sequences on the opposite tile. In the figure, this means that

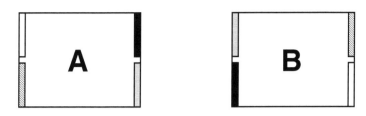

FIGURE 8.11: The two tile types for DNA self-assembly. The shaded edges denote sticky ends that bind to complementary sticky ends on the opposite tile with the same shading.

the black region on tile A will only bind to the black region on tile B, and so on. Once these tiles are fabricated, they can be mixed in solution and allowed to self-assemble. They will naturally assemble into a crystalline structure such as is shown in Figure 8.12.

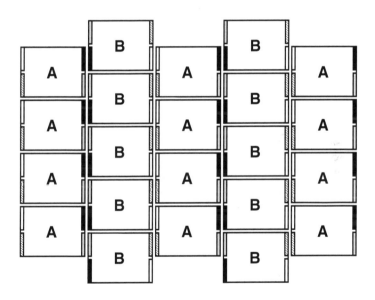

FIGURE 8.12: A two dimensional crystalline assembly of DNA tiles.

In their original study [141], the group also constructed a richer tile set consisting of four unique tile types. Using this set, they self-assembled crystalline lattices like the one shown in Figure 8.12, but with a longer periodicity. With this study, the group had taken the first step towards implementing computation in a DNA self-assembly environment. Note that Winfree et al. were *not* attempting to replace electronic computation with their DNA tiling scheme. Rather, the connection to computation allowed them to open a new doorway to control over self-assembly. With their tile sets, they had demonstrated that the construction of periodic two dimensional nanoscale lattices could effectively be programmed. This promises an unprecedented level of control over the structure of matter. The group speculated that by "decorating" tiles in the tile set with other nanoscale objects such as chemical groups, catalysts, polymer strands, or metallic nanoclusters, a wide range of nanostructured materials was within reach.

In 2004, Paul W.K. Rothemund, Nick Papadakis, and Erik Winfree took another step towards the goal of implementing computation using designed DNA tile sets [109]. To understand their approach, we need to return to Rothemund's capillary driven computing tiles and reexamine the concept of a cellular automaton.

A cellular automaton can be understood quite easily. Imagine we have a strip of squares and that each of these squares can be in one of two states. We can denote these states by colors, say white and shaded, or by digits, say 1 and 0. The state of our strip might resemble the bottom row of Figure 8.13. Now, imagine that our strip can be in different states at different instants in

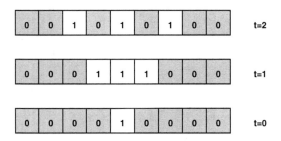

FIGURE 8.13: The evolution of a one dimensional cellular automaton.

time. Our bottom strip in the figure represents the state of our automaton at time $t = 0$. To get to time t=1, we evolve our strip according to some predefined rule. For example, imagine our rule says that each square should check the state of itself and its two neighbors and update its state according to what it finds. Say, if all three squares are shaded, the square remains shaded, if all three are white, it remains white, but in any other case the square changes its state to shaded. The time evolution of our initial string according to these rules is shown in the figure. Remarkably, this simple rule can encode a complex structure. If we evolved the system forward in time for many steps, we would produce the structure known as a *Sierpinski Gasket*.

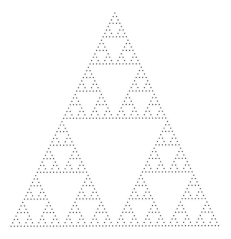

FIGURE 8.14: The Sierpinski Gasket.

This is shown in Figure 8.14. For clarity, we've used black dots to represent the one's in our array and left all else blank. Again, we emphasize the point – the complex Sierpinski structure is *encoded* in the combination of our rule set and tile labels.

There is an alternate, equivalent way, to encode the Sierpinski Gasket using the logical operator XOR. This is in fact what Rothemund had done with his capillary bond tile sets. To see this, imagine we specify our automata a little differently. This time, we'll again begin with a simple row of squares like the bottom row of Figure 8.15. But, instead of applying our rule above, we'll

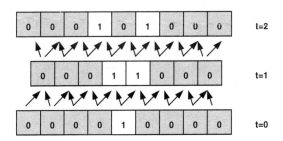

FIGURE 8.15: The evolution of a one dimensional cellular automaton implemented as an XOR operator.

simply apply the XOR operator to each pair of squares in our row and place the resulting output in a square above that is centered on the edge of the prior two squares. The result of this type of rule is shown in Figure 8.15. If we continue in this way, we again obtain the Sierpinski Gasket structure. That is, the Sierpinski Gasket is encoded in this process.

Now, a Sierpinski Gasket is easy enough to construct by hand and even simpler to construct using a computer. But, Rothemund et al. wanted to make a Sierpinski Gasket self-assemble. In order to do so, they designed a set of four DX tiles like those used by Winfree et al. in the crystallization studies above. Abstractly, these tiles were very much like the four tiles constructed by Rothemund in his capillary driven studies. From the comments above, we see that Rothemund's implementation of the XOR operator in Chapter 6 encoded the Sierpinski Gasket. But, there was a problem with using the capillary bond to form this structure. Recall that when designing his tiles, Rothemund had to use a complicated wetting pattern combined with a complex tile geometry to encode the XOR operator. While his experiments did produce limited results, this very nonspecific binding led to a high error rate. It is here that the power of DNA shines through. By using DNA tiles, binding rules can be implemented on the sticky ends with high specificity. Where the simple hydrophobic/hydrophilic alphabet was not rich enough to easily allow for such specific binding, the DNA alphabet was. To start the assembly process, the group used single long strands of DNA to encode the initial bit string. When the DNA tile set designed by Rothemund et al. attached to this string it did indeed self-assemble into a Sierpinski Gasket with a very low error rate. More details concerning the design of their tiles may be found in [109].

In 2003, a group led by John H. Reif showed that the idea proposed above of decorating tiles could be used to make functional nanostructures. In particular, this group showed how to make nanoscale protein arrays and conductive nanowires [145]. The group used the basic tile idea outlined above, but designed their tile in the shape of a cross. Their tile is shown in Figure 8.16 (A). On each of the four ends of the cross, labelled N, S, E, and W, the group placed sticky ends. As usual, the nucleotide sequences along the sticky ends could be tailored to produce different interactions and ultimately different lattice structures. When self-assembling their lattices, the group deposited tiles onto a mica substructure. They found that the presence of this substructure could modify the structure of their programmed lattices. The group was able to construct two distinct lattice types. These are shown in Figures 8.16 (B) and (C). The first type is a nanoribbon. These were long regular structures three tiles wide. The second type was a nanogrid. These were square repeating structures of tiles with a repeating corrugated design.

But, the Reif group did not stop at simply producing crystalline structures. Rather, they demonstrated that these structures could be made functional. As was suggested earlier, it is possible to "decorate" DNA tiles. Here, the group attached a molecular structure to the center of each tile. This structure, called a biotin group, selectively binds to the protein streptavidin. Once the template was constructed, streptavidin could be added to the solution and would selectively bind to the lattice, producing a regular uniform pro-

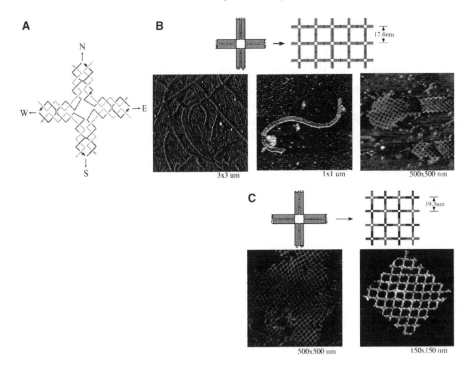

FIGURE 8.16: The Reif group's functional self-assembled nanostructures. Part (A) shows the basic tile type. (B) shows the ribbon structure formed from tile subunits, and (C) shows the grid structure formed from tile subunits. The photographs are AFM pictures of assembled structures. From Yan, et al., Science, v. 301, pp. 1882-1884, (2003), Reprinted with permission from the AAAS.

tein array. An AFM image of this self-assembled protein lattice is shown in Color Plate 11.12. In a second part of their study, the group metallized their nanoribbons with silver. This yielded a highly conductive set of nanowires. This work clearly demonstrated the practical potential of DNA tile assemblies. The ability to program DNA tiles coupled with the ability to decorate them in a functional way is a promising route to true molecular nanotechnology.

8.5 DNA Barcodes

A palindrome is a word that reads the same forwards as it does backwards. In a single strand of DNA, the presence of a palindromic pair in the nucleotide sequence allows the creation of *hairpin loops*. In 2003, another group led by John H. Reif used hairpin loops to construct a nanoscale DNA barcode [146]. Their work illustrates the potential of using a nucleating center, or seed crystal, to build a larger more complex structure. Like the tile assemblies above, it also illustrates the potential of programmable self-assembly.

In the Reif group's work, hairpin loops were used to represent the information in a barcode-like structure. Such information can be encoded in a simple bit string. The group encoded the bit strings, 01101 and 10010, by using the presence of a hairpin loop to represent a 1 and the absence to represent a 0. As mentioned above, hairpin loops are a structure that occurs in single stranded DNA with the proper nucleotide sequence. An example of a hairpin loop is shown in Figure 8.17. For nucleotides, palindromic means that a sequence is

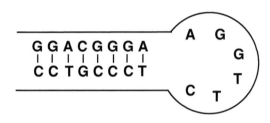

FIGURE 8.17: A DNA hairpin loop.

the same as its complementary sequence read backwards. The sequences on the upper and lower arms of the hairpin in Figure 8.17 form such a sequence. Note that in the loop part of the pin, the nucleotides remain unbound.

The group began their construction by creating an input strand of DNA that carried the desired barcode information. This input strand served as the seed crystal in their process. Next, the group designed DX tiles, like the ones above, that would attach to the input strand in the proper locations. They used two tile types. One type was decorated with two hairpin loops. One of these loops would protrude out of the plane when a tiling assembly was completed. The other would point into the plane of the assembly. The second tile did not carry any hairpin loops; this tile represented the zeros in the bit string.

When the DX tiles were mixed with the appropriate seed crystal, they self-assembled into a larger crystalline structure. However, this crystalline structure carried the information in the original bit string forward as it self-assembled. Because the tile assemblies were relatively large, the original bit string could be read from the assembled complex using an atomic force microscope. In essence, the group had created a nanoscale display.

The ability to read the pattern of the assembled structure using an AFM was an important achievement. One potential application is to DNA computing where reading the output requires PCR amplification and gel electrophoresis. Being able to directly read the output of a computation could help make DNA computing practical. However, the importance of this work goes beyond the potential display application. The group proposed that eventually the idea of using a nucleating center containing encoded information combined with DNA tile sets could be used to create scaffolds and templates for the assembly of molecular electronic and mechanical components.

8.6   DNA Origami

In 2006, Paul W.K. Rothemund introduced yet another way to self-assemble two dimensional nanoscale patterns [125, 110]. He called his approach "DNA Origami," an appropriate term for a technique able to fold long single stranded DNA molecules into arbitrary two dimensional shapes.

To accomplish this construction, Rothemund developed a sequence of five steps beginning with an approximation to the shape and ending with a self-assembled origami figure. The first step in Rothemund's process is to approximate the desired shape using DNA double helices. This approximation gives a crude first cut at the desired shape. The double helices are aligned parallel to one another and joined together by small crossover junctions. If you imagine the DNA molecules as different length strands of spaghetti, in this step you simply lay out the strands parallel to one another to get a rough approximation of your desired shape. In the second step, this structure is "rasterized." An example is shown in Figure 8.18. You can imagine this rasterized version of the smiley face lying on top of your original spaghetti strand construction. This rasterized structure will ultimately be built from a single long strand of DNA. At this point, to give the structure rigidity, short helper strands of DNA, similar to those used by William Shih and described above, are introduced. These helper strands, or DNA "staples," attach strands of the rasterized structure together. At this point, Rothemund turned to a computer to help compute the sequence of bases along the staple strands. Eventually, the staple strand and the long rasterized strand will become a single double helix with crossover junctions to give stability. Additional steps allow Rothemund to refine the design to ensure structural stability. At the end of this design process, Rothemund is left with a pattern for a long single strand scaffold nucleotide sequence and a pattern for short staple strands. When these are synthesized and mixed in solution, they self-assemble into the target two dimensional shape. An atomic force microscope image of one such folded shape is shown in Color Plate 11.14.

Rothemund's approach generalizes the construction of DNA tiles using DX

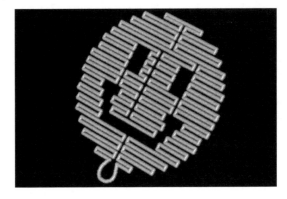

FIGURE 8.18: Folding path for Rothemund's DNA origami of a smiling face. Credit: Paul W.K. Rothemund and Nick Papadakis.

molecules. With this approach, Rothemund can synthesize a two dimensional tile with any shape. Just as with the tiles above, Rothemund's tiles can be designed to self-assemble into larger arrays. This extra level of control over the design of DNA tiles adds another layer of complexity to what can be accomplished using DNA tile based self-assembly.

8.7 DNA as a Template

In addition to being useful for building templates, strands of DNA also lend themselves to direct use as templates. In 2003, a team led by Erez Braun [69] showed that DNA could be combined with carbon nanotube technology to produce a transistor only one nanometer wide. As the group noted, prior work had established that carbon nanotubes could be outfitted with biological markers. This meant that like DNA, carbon nanotubes could be made to bind selectively. Yet, up until their study, this technique had not been used to make a functional nanostructure. To construct their transistor, the group began with a single strand of DNA. To this DNA backbone, they attached the protein known as RecA. This protein had been extracted from E. coli bacteria. Next, the group introduced a long second strand of DNA, designed so that the first strand would bind in a specified place along its backbone. Ultimately, this would allow them control over the electronic properties of their transistor. Next, the group used existing techniques to attach a second protein to a single walled carbon nanotube. This protein was chosen because of its selective binding with RecA. When the nanotubes and the DNA strand were mixed, the nanotube attached itself to the DNA in an oriented fashion. In particular, the nanotube could be aligned along the DNA backbone. Once they had the basic structure, the group deposited silver particles on the backbone. The silver also attached itself selectively, only binding in areas not protected by the RetA. Finally, the group used deposition techniques to grow gold clusters on top of the silver particles. The result was two gold coated DNA wires spanned by the carbon nanotube-DNA complex. The nanotube-DNA structure would serve as the transistor, the wires allowed electrical connections to be made.

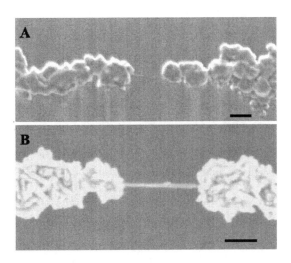

FIGURE 8.19:   The Braun group's self-assembled nanotube transistor. (A) shows an individual single walled carbon nanotube while (B) shows a rope of such nanotubes. The black bar is 100 nanometers. From Keren, et al., Science, v. 302, pp. 1380-1382, (2003), Reprinted with permission from the AAAS.

Once their device was assembled, the group probed the electrical properties of the system. They demonstrated that their device behaved like a field effect transistor. Note that the width of their self-assembled transistor was one hundred times smaller than transistors on common integrated circuits. The group had shown that self-assembly, using DNA, and integrated with carbon nanotube technology, could be used to build working electrical components. With this proof of concept, they demonstrated the feasibility of self-assembling functional electronic circuits many times smaller than the smallest circuits in use today. A scanning electron microscope image of their assembled structure is shown in Figure 8.19. For more details on their process the reader is referred to [69].

Profile - Nadrian C. Seeman

There seems to be a trend in nanoscale science. Every time someone develops a new technique for manipulating matter at the nanolevel, they immediately use that technique to write the name of their employer in really tiny letters. Nadrian C. Seeman, Ned, is the only *individual* I know of to be so honored. In a fitting tribute to the man who invented the field of DNA nanotechnology, Paul W.K. Rothemund used his DNA origami technique to write "NED" using letters only 60 nanometers tall [111].

Seeman's achievements are legendary. He is perhaps best recognized as the man who self-assembled the first three dimensional nanoscale object; the DNA cube. In fact, it was this work for which he was awarded the 1995 Foresight Institute Feynman Prize in Nanotechnology. Since then, Seeman has seemingly played a role in every major advance in the field.

Seeman did not start out as a "nanotechnologist." In fact, when he was first training as a crystallographer and biochemist at the University of Pittsburgh, the very word "nanotechnology" had yet to be coined. Yet, his work as a crystallographer was precisely what led him to develop the techniques he used to build the first nanocube. Frustrated with his inability to crystallize certain molecules, Seeman turned towards DNA for a solution. Recognizing that naturally occurring branched junctions could be made rigid by using designed DNA sequences, he quickly realized that this would not only let him build structures that would enable crystallization of his problem molecules, but would let him build practically anything he wanted. With this simple insight, Seeman had invented the field of DNA nanotechnology.

In a recent conversation, Seeman addressed the question "Why self-assembly? Why now?" His writes:

> *I've always worked with hydrogen bonded systems, which self-assemble. Always means since I was a graduate student in the late 1960's. I've been working on DNA nanotechnology (that's what it wound up being called) since the fall of 1980. So, "Why now?" only means that I haven't died yet. The other thing to say about self-assembly is that I can't think of anything on the molecular or nanoscale that doesn't self-assemble. Except in STM experiments, nobody is sitting there putting atoms or molecules together.*

Currently, Seeman is a professor in the Department of Chemistry at New York University. His group continues to focus on DNA nanotechnology, making breakthroughs at a breathtaking pace.

8.8   DNA Self-Assembly in Context

In this chapter, we've seen some of the myriad ways DNA is being put to use as a nanoscale construction material. Before concluding, we take a moment to consider DNA based self-assembly from the point of view of the last seven chapters.

First, whether we consider our particles nucleotides, DNA strands, or DNA tiles, DNA self-assembly makes use of *structured particles*. The great strength of DNA based assembly is in the complexity of the particles that can be constructed. As we saw in this section, building a Sierpinski Gasket via self-assembly using the capillary bond was hard, but using DNA tiles it could be accomplished with relative ease. The difference lies in the specific binding patterns that are readily encoded using DNA and generally difficult to encode in other systems. DNA's alphabet, used to write nature's genetic code, also provides a rich alphabet for self-assembly.

DNA self-assembly also makes use of *binding forces*. Here, it is the bonds that form between base pairs that provide the binding force. Again, the magic of DNA is the specific nature of this binding force. A's bond to T's, C's bond to G's, and they don't otherwise mix. The binding force for DNA self-assembly is highly specific.

DNA self-assembly is usually carried out in solution. This is the *environment* for this form of self-assembly. As we've seen with other systems, the interaction of particles with the environment can play a key role in the types of structures that form. The best example of that in this section is the nanoribbons built by the Reif group. The Reif group showed that it was an interaction between their tiles and the mica substrate that led to this particular pattern.

DNA self-assembly also requires a *driving force*. The particles here are nanoscale, and the process is usually carried out in solution, hence here, the driving force is random thermal agitation. This driving force does provide a means of control over the process. Changing the temperature controls the speed at which objects assemble and high temperatures increase the rate at which bonds are randomly broken. At sufficiently high temperatures, self-assembled DNA structures melt. The self-assembly process can also be controlled by manipulating this driving force. In the tetrahedra experiments of Goodman et al. we saw that rapid cooling of the solution was a crucial part of their assembly process. That is, by changing an environmental variable, temperature, they could manipulate the binding force and affect the path of assembly.

DNA self-assembly makes use of nature's other techniques as well. The principle of energy minimization dominates the design of DNA structures. Here, a good example is Chen and Seeman's nanocube. In order to induce DNA to make branched junctions, the junction state must be energetically more favorable than other accessible states. Note that there is a competition in energies in this example. Junctions bend, this takes elastic energy. Chemical potential is reduced when binding occurs, but in branched junctions this must be balanced against an increase in elastic energy. Further, the energy landscape in DNA self-assembly is often littered with local minima. These local minima present an obstacle to successful assembly and can dramatically affect the yield of a given process.

The phenomenon of nucleation also plays a role in DNA self-assembly. The clearest example of this is the barcode system designed by the Reif group. This was tile based assembly, but it was also nucleated self-assembly. The nucleation point was precisely the point that allowed the Reif group to insert their program into the system. If they nucleated with the bit string 01101 they obtained one result, if they nucleated with the bit string 10010 they obtained another.

Templates are also used in DNA self-assembly. On the one hand, DNA tile assemblies may be built and used as templates for other structures. The protein arrays and nanoscale wires of the Reif group demonstrate the feasibility of this approach. On the other hand, DNA itself may serve as a template for the construction of nanostructures. The nanoscale transistor designed by the Braun group well illustrates this approach.

The *forward*, *backward*, and *yield* problems first introduced in Chapter 5 may all be found in DNA based self-assembly. Fortunately, our knowledge of base pair binding allows some measure of success with the forward problem. Winfree's group was able to design DNA tiles that they knew would self-assemble into crystalline structures. They could solve the forward problem. But, nature always has surprises in store. Recall again the cross shaped tiles of the Reif group. Their assembly into nanoribbons was an unanticipated side effect. Sometimes when you think the forward problem is solved, nature fools you. The yield problem also arises in DNA self-assembly. No matter how carefully a system is designed, errors will occur during binding. Fortunately, the high specificity of base pair binding reduces these errors to a manageable level. But, errors are still present and methods to refine the products of DNA self-assembly still necessary. In addition, the presence of local minima in the energy landscape of a self-assembling DNA based system can lead to low yield processes. The dramatically low yield of processes to self-assemble cubes and other polyhedra illustrates this fact. Fortunately, efforts by groups such as Goodman et al. have shown possible ways to overcome the yield problem. As always, the backward problem is the most difficult. The design of the self-assembled Sierpinski Gasket and Rothemund's origami showed us two ways to attack the backward problem. The challenge now is to push these approaches to ever more complex and intricate structures.

Finally, in this chapter, we've seen several examples of self-assembling systems that fit the definition of *programmed self-assembly*. Recall that in Chapter 1, we defined programmed or programmable self-assembly as a subclass of self-assembly where the particles of the system carry information about the final desired structure or its function. This definition, like all nonmathematical definitions, is open to interpretation. An extreme point of view would be that all of the examples of self-assembly discussed in this book fit this definition. There is some merit in that argument. But, with the examples of this chapter, it begins to become clear that there is a difference between programmable self-assembly and other forms of self-assembly. DNA tiles demonstrate this most clearly. As we saw with the Sierpinski Gasket, the final structure was encoded in the tile types. The tiles performed the computation on an input bit string. By switching tile sets or input strings different structures are attainable. This begins to approach the idea of programming. One can imagine having a universal tile set and a language by which to choose the necessary tiles in order to assemble a give structure. This is what is really meant by programmed self-assembly; the systems here approach that more closely than any other system we've discussed thus far.

8.9   Chapter Highlights

- DNA, or deoxyribonucleic acid, carries nature's genetic code. It also serves as an excellent self-assembling nanoscale construction material.

- Through the use of *sticky ends* and *branched junctions*, DNA may be made to assemble into a variety of shapes including cubes, knots, links, and other polyhedra.

- The DNA *double crossover molecule* (DX) can be used to add structural rigidity to DNA constructions. It can also be used to build *DNA tiles*.

- The highly specific binding of DNA combined with DNA tiles can be used to self-assemble two dimensional crystalline structures. These structures may be periodic or aperiodic. The design of the tiles encodes the final structure in a form of *programmed self-assembly*.

- *DNA barcodes* are an example of nucleated self-assembly. Combined with DNA tiles, they offer a promising route to *programmed self-assembly*.

- *DNA Origami* is a method for self-assembling arbitrary two-dimensional structures from a single strand of DNA aided by short helper strands.

- DNA can serve as a template for self-assembly in two ways. First, DNA tiles can be assembled and made functional. In this way, other structures can be built using the tile assembly as a template. DNA may also serve directly as a template.

- DNA self-assembly makes use of nature's four key components, *structured particles*, *binding forces*, an *environment*, and a *driving force*. Further, the *forward, backward,* and *yield* problems all present challenges in DNA self-assembly.

8.10   Exercises

Section 8.2

1. Consider a fictitious DNA molecule that makes use of only two bases. Let's call these bases 0 and 1 and assume that $0-1$ bonds can occur but not $0-0$ or $1-1$. For this encoding, what size groups of bases would be needed to specify all twenty amino acids uniquely?

2. For the fictitious DNA molecule of the last problem, show how to design a branched junction. Exhibit sequences that lead to both movable junctions and fixed junctions.

Section 8.3

3. Construct a sequence of base pairs for real DNA that allows one to build a three armed junction. Choose your sequence so that the junction is fixed.

4. It is possible to build junctions that have more than four arms. Show how to build a five armed junction.

Section 8.4

5. Write a simple computer program to construct the Sierpinski Gasket.

6. Many other rule sets are possible for a cellular automaton. In the Related Reading section, there is a pointer to Stephan Wolfram's classification of possible rule types. Pick a rule type and implement it on a computer.

7. For the rule type you picked in the last problem, design a set of DNA tiles that would implement this rule type.

8. Consider the cross-shaped tiles of this section. Suppose the tiles were designed with $NNSS$ edges instead of $NSEW$. What structures would you expect to appear? If you used a single tile with $NSEW$ edges and a mixture of tiles with NN labels only, what structure would you expect to appear?

9. The Sierpinski tile set implemented the logical operator XOR. How would you design a tile set to implement the operator OR? If this operation is applied to the bit string of this section, what structures would appear?

Section 8.5

10. Discuss the template structures of this section as compared to the template structures of Chapter 6. How are they alike? How do they differ? Which methods are the most powerful?

Section 8.6

11. How does DNA Origami compare with the magnetic origami of Chapter 5? Discuss this question in terms of energy minimization and the yield problem.

Section 8.7

12. How does the use of DNA as a template compare to the template self-assembly methods of Chapter 6? What advantages does DNA offer over other methods?

8.11   Related Reading

Watson's entertaining account of the discovery of the structure of DNA is a classic "must read."

J.D. Watson, *The Double Helix*, Penguin Books, 1969.

Ned Seeman's Scientific American article on DNA and nanotechnology is a great introduction to the area.

N.C. Seeman, *Nanotechnology and the Double Helix*, Scientific American, 290 (2004).

Turberfield's article in Physics World outlines the engineering properties of DNA.

A. Turberfield, *DNA as an Engineering Material*, Physics World, 16, (2003).

Stephen Wolfram is the world's expert on cellular automata. You can learn everything you ever wanted to know about them from his text.

S. Wolfram, *A New Kind of Science*, Wolfram Media, 2002.

8.12 Notes

1. The fascinating story of Watson and Crick's discovery is recounted in a book by Watson. The book is listed in the Related Reading section for this chapter. It is a classic of scientific literature and a definite must read.

Part III

The Future

Chapter 9

Models of Self-Assembly

Profound study of nature is the most fertile source of mathematical discoveries.

Fourier, *Analytical Theory of Heat*

9.1 Introduction

One does not have to understand the science of optics to appreciate the beauty of a rainbow; but it helps. In the same way, one need not master the mathematics of self-assembly in order to appreciate the power of the concept; but here too, it helps. In the first two parts of this book we focused on descriptions of self-assembling systems. At times, we made use of mathematics, but ultimately our focus was on experiment rather than theory. In this chapter, we shift our focus and examine the various theoretical approaches to understanding the phenomenon of self-assembly.

There are as many different approaches to mathematically modelling self-assembly as there are examples of physical self-assembling systems. In the end, the type of model one constructs depends upon the type of question one wishes to answer. These questions can vary wildly. At one end of the spectrum, we have models built to illuminate the behavior of one specific self-assembling system. Such a model can have great utility. If accurate, it can help reduce the number of costly or time consuming experiments one needs to conduct. It can clarify the role of various parameters in the system and give a picture of parameter space that might otherwise be inaccessible. At its best, it can clarify a complex situation, help guide experiment, and identify new experimental regimes to be explored. At the other end of the spectrum we have abstract models of the phenomenon of self-assembly. These models are usually divorced from any particular experimental system; rather they seek to capture the behavior of some large class of self-assembling systems. These models too, can have great utility. At their best, they can help us answer "What is possible?" types of questions. Is it possible to self-assemble a Sierpinski Gasket in a system containing only two tile types? Is it possible to self-assemble a cell given infinitely many tile types? These are the types of questions that abstract models are best at answering.

However, there is no hard and fast boundary between these types of models. Models of a particular physical system are often found to apply to other systems, systems that at first might seem unrelated. These models are perhaps more abstract than we initially thought. Abstract models take their inspiration from physical systems and in seeking to capture general principles, often end up capturing real behavior remarkably well. At times, a model that initially seemed abstract may end up being physically realizable, and end up showing us a new route to self-assembly.

Nor is there any mathematical distinction between these types of models. The equations of continuum mechanics can help us develop a detailed description of the shape of a meniscus, but they can also be implemented in a computer, governing the behavior of fictitious particles that have no counterpart in the real world. Seemingly pure branches of mathematics, such as graph theory, which lends itself nicely to several abstract approaches, also lends itself nicely to robotic control schemes for real world engineered particles. Similarly, computer simulation plays an important role in the analysis of every kind of model. Both physically driven models and abstract models have a tendency to become analytically intractable. In both cases, numerical simulation becomes a necessity.

Nonetheless, for clarity in the discussion, we will make a distinction between these two types of models. We'll divide this chapter into two main sections. In the first, *Physical Models*, we'll describe approaches that stay close to one physical system or some small subclass of physical systems. In the second, *Abstract Models*, we'll examine approaches to "What is possible?" type questions.

In Section 9.2, *Physical Models*, we begin with a mathematical model of the structured surfaces discussed in Chapter 6. This model asks the question: What can be accomplished if an electric field is used to manipulate the minimal energy surfaces of Chapter 6? This model is very much at the "single experimental system" end of the spectrum. Through this model we'll see how key parameters in a problem may be identified and how a model can help us understand parameter space and suggest experimental directions. Next, we'll examine a model that attempts to explain why the helix is such a familiar structural motif in nature. In contrast to the structured surface model, this model focuses on a class of self-assembling systems rather than on a single experimental setup. We'll see how such a model can be useful, both to give insight into a broad problem, and to actually predict experimental results. For our third model, we'll return to the first system we discussed in Chapter 6: the self-assembling tile system of Hosokawa et al. The model we'll discuss is drawn from their original paper [62] describing their experimental and theoretical results. We'll see how a model inspired by chemical reaction kinetics can capture the behavior of a tile based self-assembling system. Finally, in this section, we'll discuss the so-called *waterbug model*, due to Eric Klavins.

This final model is again unattached to any particular physical system, but is inspired by a class of such systems. With this model, we'll see how theory can aid in the design of physical systems.

In Section 9.3, *Abstract Models*, we focus on three abstract approaches to modelling self-assembly. The notion of a *conformational switch* is the focal point of the first of these models. We encountered conformational switching in Chapter 3 when we discussed the tobacco mosaic virus. We also encountered this notion when we discussed proteins and again in part two of this book in the context of several different engineered systems. The model of conformational switching presented in this section attempts to characterize the power of a conformational switch to encode for a given assembly sequence. The second model we consider is based on the notion of a graph grammar. This model generalizes the conformational switch model and within the context of the model is able to provide a constructive solution to the backward problem of self-assembly. The final model we consider is the *Tile Assembly Model*. This important model provides the link connecting self-assembly and computation. We'll see how this model has been used to explore the question of complexity of a self-assembling system and how this model provides a promising route to programmed self-assembly.

One final note before we begin – to understand the details of every model discussed in this chapter requires a broad mathematical background. Here, we won't focus on these details. Rather, we'll attempt to provide a sense of the thinking behind the model, the questions it seeks to address, and the importance of the answers to those questions. Further, where it seems most appropriate, we'll fill in the mathematical background needed to understand the basics of the model. However, this may not always be enough. If you find the details of a particular model in this chapter to be confusing or inaccessible, skip them. You should still be able to get a sense of the model. If you still find a particular model to be heavy going, skip it entirely. The subsections in this chapter are mostly independent.[1] I encourage you to find a modelling approach and a set of questions that excites you, and to continue from there.

9.2 Physical Models

In this section, we present four models that are focused on one particular physical system, or on a small subclass of such systems. We've chosen these models as representative examples. Many more such efforts exist. We refer the reader to [12, 64, 88, 148] for information on similar approaches to modelling self-assembly. Further, if you return to part two of this book, you'll find that the references to the experimental systems discussed often contain models of these systems.

9.2.1   Modelling Structured Surfaces

In Chapter 6, we examined an approach to self-assembly, pioneered by the Whitesides group, that we called *structured surfaces*. Recall that in this approach, droplets of PDMS were placed between rigid plates and that by changing the wettability of these plates, the density of the surrounding fluid, and the orientation of these plates, the shape of the surface assumed by the PDMS droplet could be controlled. In turn, the PDMS could be cross-linked, or solidified, and hence objects with interesting shapes constructed without the use of a mold or template. In their original article, the Whitesides group [70] also conjectured that electric or magnetic fields could be used to obtain an additional level of control over the shape of these surfaces. The model of this section explores this idea. This model is due to Derek Moulton; further details may be found in [93].

To begin, Moulton replaced the PDMS droplet of Whitesides by a soap film spanning two identical concentric rings. Working with a soap film allowed him to remove the volume constraint inherent in the PDMS system and work with simple boundary conditions at the endpoints. Additionally, Moulton was able to carry out experiments with the system he devised. As we discussed in Chapter 6, a soap film spanning two rings naturally forms a catenoid. To see how this shape could be manipulated, Moulton added an outer electrode to the system surrounding the two ring soap film structure. By applying a voltage difference, an electric field could then be created in the gap between the soap film and the electrode. The basic setup of this system is shown in Figure 9.1.

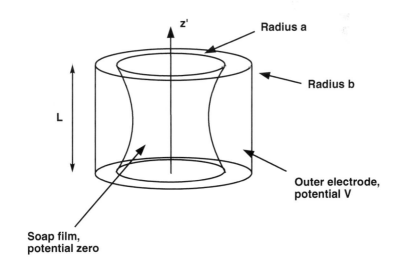

FIGURE 9.1:   The geometry of the soap film and electrode system.

By adding an electric field, Moulton added a second energy to the problem. The soap film will attempt to minimize its surface energy, we saw this in Chapter 6. But, there is also energy stored in the electric field. The shape selected by this system will minimize the total energy, i.e., the sum of the surface and field energies.

To derive an equation that would predict this shape, Moulton first needed to derive expressions for each of these energies. In sketching this derivation, we use the notation of Figure 9.1. Note that in the figure, the shape of the surface is specified by the function $u'(z')$ and the electric field is specified in terms of the potential function $\psi'(r', \theta, z')$. The potential is assumed constant on the outer electrode and on the soap film. The rings supporting the film are of radius a, they are placed a distance L apart, and the outer electrode has radius b.

In electrostatics, the electric field, $\vec{E}$, can be specified entirely in terms of the potential function through $\vec{E} = -\nabla \psi'$. Since in the absence of free charges, the field satisfies

$$\nabla \cdot \vec{E} = 0 \tag{9.1}$$

the potential, ψ', satisfies the Laplace equation

$$\nabla^2 \psi' = 0. \tag{9.2}$$

The fixed potential conditions on the soap film and outer electrode translate into the boundary conditions

$$\psi'(b, \theta, z') = 0 \tag{9.3}$$

$$\psi'(u'(z'), \theta, z') = 0. \tag{9.4}$$

Now, note that the soap film surface, $u'(z')$, must also satisfy boundary conditions. In particular,

$$u'(L/2) = u'(-L/2) = a. \tag{9.5}$$

Notice that we already have four parameters in this problem, a, b, L, and V. In models such as this, it is convenient to introduce nondimensional variables. This not only simplifies the discussion, but also helps one uncover the relative importance of various terms in the model and helps reduces the dimension of parameter space to a manageable level. This process is called *nondimensionalization*. You can find a thorough explanation of this process in [99]. Here, we introduce the variables

$$z = \frac{z'}{L}, \qquad r = \frac{r'}{b-a}, \qquad \psi = \frac{\psi'}{V}, \qquad u = \frac{u'}{a}. \tag{9.6}$$

With these substitutions, Equations (9.2) through (9.4) become

$$\frac{\partial^2 \psi}{\partial r^2} + \frac{1}{r} \frac{\partial \psi}{\partial r} + \epsilon^2 \frac{\partial^2 \psi}{\partial z^2} = 0 \tag{9.7}$$

$$\psi = 1 \quad \text{at} \quad r = \frac{b}{b-a} \tag{9.8}$$

$$\psi = 0 \quad \text{at} \quad r = \frac{a}{b-a}u(z). \tag{9.9}$$

This set of equations is in non-dimensional form. The key dimensionless parameter that arises here is $\epsilon = (b-a)/L$. Physically, ϵ is an aspect ratio, comparing the size of the gap to the length of the device. In his experimental system, Moulton found that ϵ was in fact a small parameter. This fact will be used to simplify the analysis below.

Now, the energy stored by the electric field is given by a volume integral taken over the region between the soap film and outer electrode. In particular,

$$\text{Electrostatic Energy} = -\frac{\epsilon_0}{2}\int |\vec{E}|^2 = -\frac{\epsilon_0}{2}\int |\nabla\psi'|^2. \tag{9.10}$$

Notice that to compute this energy, we need to know ψ' or equivalently ψ. This means that we need to solve Equations (9.7) through (9.9). However, this is not easy. First, the shape of the domain is not regular, and second, the shape of the domain is not even known; it depends on $u(z)$. But, an approximate solution for ψ can be obtained by exploiting the fact ϵ is a small parameter. This requires *asymptotic analysis*; we'll skip the details. Given this approximate solution, an expression for the electrostatic energy can be obtained. It can be further simplified by using the divergence theorem[2] and the boundary conditions. At the end, we obtain

$$\text{Electrostatic Energy} = -\pi\epsilon_0 V^2 L \int_{-1/2}^{1/2}\left(\log\frac{\delta}{u(z)}\right)^{-1}dz. \tag{9.11}$$

Note that here, the dimensionless parameter δ is the ratio b/a of the radii of the outer and inner cylinders.

With the electrostatic energy in hand, we are halfway there. We still need an expression for the surface energy in the problem. But, we know that this energy is simply proportional to the change in surface area, and we already saw how to compute this in Chapter 6. In terms of our nondimensional variables, we can write the surface energy as

$$\text{Surface Energy} = 2\pi T L a \int_{-1/2}^{1/2} u\sqrt{1+\sigma^2 u_z^2}dz. \tag{9.12}$$

Here, the subscript on the u denotes differentiation with respect to z, the dimensionless parameter σ equals a/L, and T is the film tension.

Finally, forming the sum of our two energy expressions and dividing by $2\pi T L a$ we obtain an energy functional for our system

$$E[u(z)] = \int_{-1/2}^{1/2}\left(u\sqrt{1+\sigma^2 u_z^2} - \frac{\lambda}{\log(\delta/u)}\right)dz. \tag{9.13}$$

Here, the dimensionless parameter λ is given by

$$\lambda = \frac{\epsilon_0 V^2}{2Ta}. \tag{9.14}$$

This parameter measures the relative strengths of the electrostatic and surface energies in our system. This function, $E[u(z)]$, is not so far from other energy functionals we've encountered in this book. It maps the shape of our surface, $u(z)$, to a real number denoting the energy of the system. As usual, we claim that nature chooses the shape that makes this energy as small as possible. Fortunately, this functional is in a form such that the Euler-Lagrange equation introduced in Chapter 6 can be applied. Doing so, we find that $u(z)$ satisfies

$$\frac{1 + \sigma^2 u_z^2 - \sigma^2 u u_{zz}}{(1 + \sigma^2 u_z^2)^{3/2}} = \frac{\lambda}{u \log^2(\delta/u)} \tag{9.15}$$

plus the boundary conditions

$$u(1/2) = u(-1/2) = 1. \tag{9.16}$$

Now, a complete and detailed analysis of Equations (9.15) and (9.16) is beyond the scope of this book. These details are developed further in the exercises and can be found in [93].[3] Rather, here, let's make a few observations about this model, about how it can be used, and why such a model is important in self-assembly.

First, note that the left hand side of Equation (9.15) is actually the negative of the mean curvature operator we encountered in Chapter 6. Hence, Equation (9.15) can be rewritten as

$$Hu = \frac{-\lambda}{u \log^2(\delta/u)}. \tag{9.17}$$

This is interesting to note because this represents a generalization of the standard equation of constant mean curvature surfaces. In fact, Moulton calls his surfaces *field driven mean curvature* surfaces. Self-assembly has given us a new and interesting problem in mathematics. More importantly, note that this formulation has greatly simplified and clarified the parameter space of the original problem. Here, we find three nondimensional parameters, λ, σ, and δ. The parameter space of the original formulation was six dimensional. With this formulation we see that not every parameter independently effects the shape of the surface. Rather, it is the ratio of groups of these parameters that is important. Next, we can immediately see that there are two special exact solutions to this problem. The first occurs when no voltage is applied; in this case the surface assumes the shape of a catenoid. The second is a cylindrical solution, $u = 1$, that occurs when exactly the right voltage is applied. This already begins to give some insight into the control that can be obtained over surface shape via the application of an electric field. Further, using *perturbation theory*, these special solutions can be used to construct approximations to nearby solutions. In this way Moulton was able to answer a rather interesting question. In particular, he knew that in the absence of applied voltage, there was a critical value of σ, such that the catenoid solution disappeared when this critical value was passed. We saw this in Chapter 6. Moulton asked whether or not an applied voltage would allow one to self-assemble nearly catenoid shaped surfaces beyond the critical value of this parameter. His analysis yielded an affirmative answer to this question. Yet, to obtain such surfaces, parameter values must be balanced very carefully. It is unlikely that this parameter regime would be found by experiment alone.[4] Further, through analysis of this model, Moulton was able to show that there are limits to what the applied field can do. He showed that as the voltage is increased, the soap film begins to bulge outwards towards the outer electrode. But, it does not continue to do so in a smooth way until it reaches the outer electrode. Instead, there is a critical value of the applied voltage beyond which the soap film simply pops. Finally, we should note that Moulton has carried out several experiments with this and related systems. He has obtained good agreement between his theory and experimental results.

This type of model is important in self-assembly precisely because it helps guide the experimenter through a large and treacherous terrain. To be truly successful, models such as these must be tightly coupled to experimental efforts. The development of virtually every self-assembling system discussed in this book can benefit from models such as these. The models may not take the form of the model discussed here; energy minimization may not apply, or such an approach may be too difficult to be of use. But, tight coupling between experimental efforts and theoretical efforts such as these promises to help us push the boundaries of experimental self-assembly rapidly forward.

9.2.2   Modelling Helix Formation

In the cell, DNA naturally assumes a helical shape. This basic design reoccurs throughout nature. In Chapter 3, we saw that the secondary structure of proteins consisted of α helices and β sheets. The α helix is, obviously, another example of helix formation in nature. The β sheet is yet another example, consisting of a sequence of helices lying side by side. Helix formation in proteins is one restricted example of the general protein folding problem discussed in Chapter 3. In this section, we consider a model due to Yehuda Snir and Randall D. Kamien [126] that asks why the helix design is so prevalent in nature. Basically they asked: Why do natural structures so frequently self-assemble into a helix?

Their model, while still in our class of physical models, is more abstract than the model above. Rather than attempting to understand helix formation in a particular protein, or other particular physical system, Snir and Kamien attempted to find minimal conditions under which helix formation would occur. They began by considering a long cylindrical rod of radius t

immersed in a solution. They also imagined that this solution contained some concentration, n, of hard spheres of radius r. Next, they posited a simple interaction mechanism between their spheres and their rod. They imagined their rod was surrounded by an annular region that was inaccessible to the hard spheres. Next, they considered the entropy of the spheres. If the inaccessible region were very large, the spheres would be confined to some small region of space. In turn, this would imply that their entropy was low. If this region were smaller, the spheres entropy would increase. Now, fixing the excluded volume, they required that the system attempt to maximize the entropy of the hard spheres. With the excluded volume fixed, the only way entropy could increase was if the rod were to bend or fold. Folding resulted in increased entropy because it created regions where the inaccessible annular region overlapped itself. This reduced the inaccessible region seen by the spheres and hence increased their entropy. But, there is a cost to this folding, increased elastic energy. That is, it requires more elastic energy for the rod to bend than for it to stay straight. They required that their system minimize this elastic energy as well as maximize the entropy of the hard spheres. It was a balance of these two that they speculated might lead to helix formation.

This balance of entropy and energy can be expressed in the *free energy* for the system. For the Snir and Kamien system the total change in the free energy due to bending of the rod is given by

$$\triangle F = \frac{1}{2}Ll_p\kappa^2 - nV_0. \tag{9.18}$$

Here, L is the length of the rod, κ is the curvature of the rod in a helical formation, l_p is the persistence length and measures the stiffness of the rod, n is the hard sphere concentration, and V_0 is the reduction in the excluded volume that occurs when the helical shape is assumed.

Notice that in their model, Snir and Kamien assumed a helical formation for the rod. A helix has constant nonzero curvature, hence the κ term in Equation (9.18) is a constant. This model could be generalized by allowing the rod to assume any shape. In this case the κ^2 term would be replaced by an integral of κ^2 over the length of the rod. Since the shape of the rod can be specified completely in terms of its curvature, this would turn Equation (9.18) into a functional, mapping the rod shape to the free energy. Unfortunately, the second term in Equation (9.18) is not so easily expressed as an integral. This means that there is no easy approach to deriving a differential equation for the rod shape. So, instead, Snir and Kamien took their cue from nature, and replaced the unmanageable space of all possible rod shapes, with the manageable space of helices. Also note that Equation (9.18) embodies the competition between elastic energy and sphere entropy discussed above. The first term on the right is always positive and captures the increase in free energy needed to bend the rod. The second term on the right is always negative and captures the increase in entropy of the spheres that occurs when the overlap region becomes larger. This second term is also proportional to the concentration of hard spheres. As this concentration increases, a small change in the overlap volume results in a large change in the entropy of the particles.

The difficult aspect of analyzing this model is in computing the overlap volume for a given rod configuration. Restricting attention to helices made this task manageable, but numerical computations were still necessary. Performing these computations, the team then characterized their results in terms of three dimensionless parameters. The first dimensionless parameter, $c = P/R$, allowed them to specify their helix in terms of its pitch, P, and radius, R. The pitch of a helix is the distance between successive turns. The second dimensionless parameter, r/t, allowed them to characterize the relative size of the hard spheres as compared to the radius of the rod. The final dimensionless parameter, $\theta = nr^3/(l_p/t)$, serves as a control parameter, allowing them to compare a reference entropy with a reference energy.

Now, if the parameter r/t is fixed, there is a single value of c that maximizes the overlap volume. Hence, c may be regarded as a function of r/t. Numerically, the group computed this functional dependence and showed that as r/t went to zero, c tended towards a limiting value, $c^* \approx 2.5122$. This meant that for small spheres, a helix would form with this given pitch to radius ratio. Remarkably, this ratio compared well with measured ratios found in helical proteins, where the lower bound $c \approx 2$ had been found. The group also plotted θ as a function of r/t. This gave them a picture of the configuration space of the helix. For low values of θ, the tube forms a stretched helix, but as some threshold value is crossed, the helix collapses into a tightly wound spiral.

To illustrate the role of lattice models and also to gain more insight into the work of Snir and Kamien, let's examine a simplified lattice model version of their energy-entropy helix formation system. To begin, we'll construct our "rod" on a lattice like those discussed in Chapter 3. We can imagine specifying the configuration of our rod by starting at the origin, picking a direction, moving one step in this direction, and then repeating the process. At each step, we form an edge, or a bond. A sample rod is shown in Figure 9.2. If we don't allow our rod to fold back onto itself, this means that in the first step we have four directions to choose from, but in subsequent steps, only three. We can specify our rod in terms of a vector

$$\vec{m} = [m_0, m_1, \ldots, m_N]. \tag{9.19}$$

The first component of the vector, m_0, takes the value 1, 2, 3, or 4, according as our initial step. We label a step in the positive x-direction by 1, in the positive y-direction by 2, and so on. The remaining components of the vector take on one of three possible values, -1, 0, or 1. We label a step with no turn by 0, a counterclockwise step by -1, and a clockwise step by 1. Now, we imagine that every time we bend our rod, or in this case take a step perpendicular to the last step, that it costs elastic energy. The total elastic

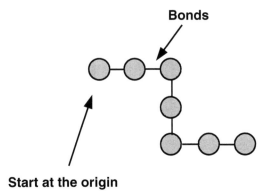

Bonds

Start at the origin

FIGURE 9.2: A bent rod in our lattice model.

energy of our rod can then be written as

$$\text{Elastic Energy} = \alpha \sum_{i=1}^{N} m_i^2 = \alpha m_b. \qquad (9.20)$$

Here, we have introduced the constant α to measure the magnitude of the energy required for one bend. The constant m_b is simply the total number of bends in our rod. Note that the direction of the bend does not matter, this is why the m_i are squared in the sum. Next, to capture the notion of excluded volume, we define V_s to be the number of empty lattice sites adjacent to our straight rod, Figure 9.3. The reader may verify that for a rod with $N + 1$ bonds, $V_s = 2N + 6$. For a bent rod, we define V_e to be the number of lattice sites adjacent to the rod. Hence, we can write an excluded volume energy as

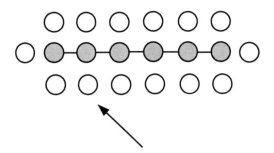

Empty circles are the excluded volume

FIGURE 9.3: The definition of excluded volume in the lattice model.

$$\text{Excluded Volume Energy} = \beta(V_s - V_e) = V(\vec{m}). \qquad (9.21)$$

Here, the parameter β may be thought of as a measure of the external concentration of hard spheres in our system. Our total energy is then

$$E[\vec{m}] = \alpha m_b - \beta V(\vec{m}). \qquad (9.22)$$

Note that this too is an energy functional. This time, it maps our shape vector, $\vec{m}$, to the real number denoting the energy of the system.

Finding the shape assumed by our rod now reduces to computing $E[\vec{m}]$, for all rods of a given length, and then picking the one with least total energy. We illustrate with a simple example. If $N = 1$, we have rods consisting of only two steps. Ignoring symmetries, there are only two possible rod shapes, the straight rod, or the right angled rod. These are shown in Figure 9.4. These

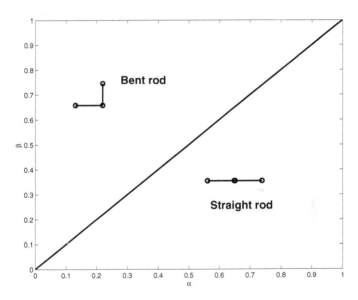

FIGURE 9.4:   A phase diagram for our lattice model, $N = 1$.

are specified by the vectors $[1,0]$ and $[1,1]$. The energies are given by

$$E[[1,0]] = 0 \qquad (9.23)$$
$$E[[1,1]] = \alpha - \beta. \qquad (9.24)$$

Hence, we immediately see that if $\alpha > \beta$ the rod remains straight, while if $\alpha < \beta$, the rod bends. This lets us sketch the phase space shown in Figure 9.4.

This lattice model, while easily grasped, suffers from a similar defect to the model above. If N becomes large, the number of possible rod shapes, or the size of the state space of the model, becomes enormous. This is a typical problem of models of self-assembly and one that is difficult to deal with.

The model of Snir and Kamien presents us with a nice example of a physical model that seeks to capture the behavior of a class of self-assembling systems. Their model provides good insight into minimal conditions for helix formation. The fact that their model predicts a pitch to radius ratio that is in good accord with experiment is remarkable. It illustrates the power of conceptually simple models to capture the actual behavior of real physical systems.

9.2.3   Chemical Kinetics Models

In Chapter 6, we examined the self-assembling tiles of Hosokawa et al. At the time, we noted that in their original paper, Hosokawa et al. presented a mathematical model of their system. In this section we return to the Hosokawa system and examine their modelling approach. We note that this model is a physical model, but relies on certain abstractions. Particle collisions are treated using chemical reaction kinetics; an assumption that may or may not be valid. Further, while Hosokawa et al. only applied their model to their self-assembling system, the basic approach is more widely applicable and has been used and extended by other authors.

Recall that Hosokawa et al. designed a set of tiles with the intent of self-assembling a simple finite cluster. In their first set of experiments, they encountered what they called *reverse coupling*. In their second set of experiments, Hosokawa et al. constructed tiles that enabled them to overcome the reverse coupling phenomenon of their first set. These tiles could exist in one of four stable cluster types. These clusters are pictured in Figure 9.5. Denoting these clusters by X_i, as shown in the figure, Hosokawa et al. proceeded by analogy with chemical kinetics and described cluster-cluster binding through a set of four reaction equations

$$2X_1 \longrightarrow X_2 \qquad (9.25)$$
$$X_1 + X_2 \longrightarrow X_3$$
$$X_1 + X_3 \longrightarrow X_4$$
$$2X_2 \longrightarrow X_4.$$

Next, they captured the state of their system at time t in the state vector

$$\vec{x}(t) = [x_1(t), x_2(t), x_3(t), x_4(t)]^T. \qquad (9.26)$$

Here, T denotes the transpose so that $\vec{x}$ is actually a column vector, and the x_i measure the amount of species X_i at time t. Next, they formulated a discrete model for their system, based on reaction kinetics, and similar to the continuous reaction kinetic based models we have explored elsewhere in this

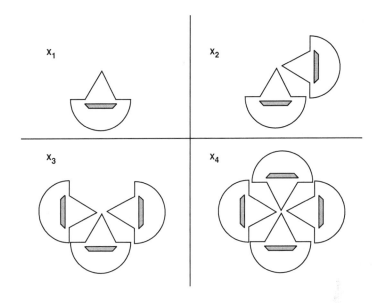

FIGURE 9.5: The four cluster types in the model of Hosokawa et al.

text. In particular, they assumed their system evolved according to

$$\vec{x}(t+1) = \vec{x}(t) + AP(\vec{x}(t)). \tag{9.27}$$

To understand this discrete dynamical system, let us proceed in two steps. For the moment, let's ignore P, and focus on the matrix A. The matrix A is given by

$$A = \begin{bmatrix} -2 & -1 & -1 & 0 \\ 1 & -1 & 0 & -2 \\ 0 & 1 & -1 & 0 \\ 0 & 0 & 1 & 1 \end{bmatrix}. \tag{9.28}$$

The components of this matrix come from the reaction equations. In particular, the ijth component, a_{ij}, is the number of X_i in the jth reaction equation. So, a_{11}, is the number of X_1's appearing in the reaction equation

$$2X_1 \longrightarrow X_2. \tag{9.29}$$

The minus sign reflects the fact that X_1 is used up in this reaction, i.e., it appears on the left hand side of the reaction equation. Now, we need to return to $P(\vec{x}(t))$. Clearly, since it is to multiply the matrix A, P must take the vector $\vec{x}$ as input and return a column vector of the same length. That is, P is itself a 4×1 vector. Now, the vector P captures the rate at which our different reactions occur. Its jth component, P_j, represents the probability the jth reaction occurs in the given time step. If we proceeded directly from the Law of Mass Action, we'd find that

$$P(\vec{x}(t)) = k[x_1^2, 2x_1x_2, 2x_1x_3, x_2^2]^T \qquad (9.30)$$

where k is a rate constant for the reactions. But, this assumes that all reactions occur at the same rate. That was clearly not what Hosokawa et al. had observed in their experiments. Hence, they proposed a modified form for P, namely

$$P(\vec{x}(t)) = \frac{1}{S^2}[P_{11}^b x_1^2, 2P_{12}^b x_1 x_2, 2P_{13}^b x_1 x_3, P_{22}^b x_2^2]^T. \qquad (9.31)$$

Here, S is the total number of clusters of all types, and P_{ij}^b is the conditional probability that a bond occurs between and X_i and X_j on the condition that they collide.

Hosokawa et al. offered a simple geometric model for approximating the P_{ij}^b. The reader may find this calculation in [62]. At this point, Hosokawa et al. turned to numerical simulation. Fortunately, their basic model is easy to implement computationally. They compared their experimental results with the output of their simulation. The agreement was not good. But, recall that in Chapter 6 when we discussed the Hosokawa system, we noted that the reaction

$$X_1 + X_3 \longrightarrow X_4 \qquad (9.32)$$

hardly ever occurred. This observation did not accord with their calculated value of $P_{13}^b \approx 0.188$. So to more accurately reflect what was observed experimentally, they set P_{13}^b to zero. This time, their model did give good agreement with their experimental results.

The model of Hosokawa et al. gives us another nice example of a theoretical model closely coupled to experimental efforts. The experiment and analogy with chemical kinetics suggested the form of the model. Experiment also helped identify the values of parameters in the model. This is a common and important form of feedback between theory and experiment. In a purely theoretical effort, the unusually low value of P_{13}^b could not have been predicted. Yet, once the appropriate parameter values were uncovered, Hosokawa et al. could use their model with confidence to predict the behavior of systems larger than those experimentally accessible.

9.2.4 The Waterbug Model

As our final example of a physical model, we discuss the *Klavins' Waterbug Model* first presented by Eric Klavins in [71]. This model aims at both capturing the behavior of and develops methods for control of self-assembling systems. The model was motivated by the many different tile based systems we examined in Chapter 6, as well as the simple Cheerios effect phenomenon of Chapter 5. Klavins noted that this model, while formulated in terms of the capillary bond, could easily be adapted to systems that use magnetic, electrostatic, or other forces as their binding force.

As the basic subunit, or particle, in his model, Klavins used a *waterbug*. This is simple cross-shaped structure reminiscent of the Reif group's tiles from Chapter 8. The Klavins waterbug is shown in Figure 9.6. Note that

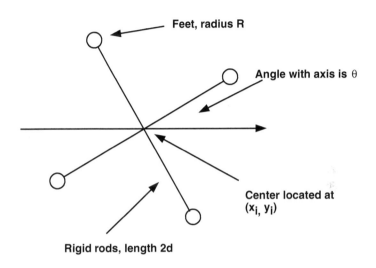

FIGURE 9.6:   The geometry of the waterbug model.

this particle consists of two equal length rigid rods joined at their centers. At the ends of each rod we find "feet." These are buoyant particles whose wettability can be controlled. As with the many tile systems of Chapter 6, these waterbugs are constrained to live on a liquid surface. The meniscus effect then creates attraction or repulsion between the feet of various waterbugs. Next, let's sketch the derivation of the waterbug model.

To capture the behavior of a system of n waterbugs, we begin by specifying the orientation and location of each particle. In particular, we define

$$q_i = [x_i, y_i, \theta_i]. \tag{9.33}$$

The components x_i and y_i give the location of the center of the bug, the component θ_i gives the angle of its rotation from the x-axis. These coordinates are shown in Figure 9.6. From the q_i, the location of each foot of a given tile can be computed. We find

$$w_{i,j} = [u_{i,j}, v_{i,j}] \tag{9.34}$$

where $w_{i,j}$ gives the position of the ith foot on the jth tile. Note that the $u_{i,j}$ and $v_{i,j}$ are specified in terms of the q_i. For example, $u_{i,1} = x_i + d\cos(\theta_i)$.

Hence, the complete state of the system is specified by the q_i. It is convenient to define

$$q = [q_1, q_2, \ldots, q_n] \qquad (9.35)$$

and use q to denote the state of the system.

To model the dynamics of the system, again, an energy approach is useful. Here, the idea is to derive an expression for the potential energy, an expression for the kinetic energy, and then to use *Lagrangian dynamics* to derive equations of motion. In previous systems, we've minimized energy to derive governing differential equations. In Lagrangian dynamics, we minimize the *action* of the system. The action is defined as the difference between the total kinetic and total potential energy in the system. The reader will find more information on Lagrangian dynamics in the related reading section.

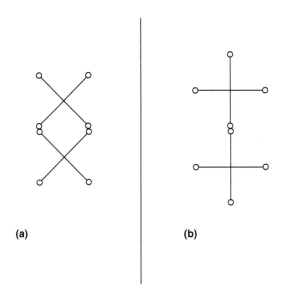

(a)   **(b)**

FIGURE 9.7: Stable and unstable configurations in the waterbug model. Type (a) is stable, type (b) is unstable.

The potential energy in this system is the sum of the pairwise potential of all possible pairs of interacting particles. The potential energy between particles i and k is given by

$$u(q_i, q_k) = -\sum_{j=1}^{4} \sum_{l=1}^{4} c_{ijkl} K_0(\rho \| w_{i,j} - w_{k,l} \|). \qquad (9.36)$$

This expression for the potential was derived by assuming that the feet interacted like spheres on the surface of a fluid. In Chapter 5, we saw that the force between such particles could be obtained approximately in terms of the modified Bessel function K_1. The potential energy can be expressed in terms of the gradient of the force, and the derivative of the modified Bessel function K_0 is the negative of K_1. This is the origin of the modified Bessel function, K_0, in Equation (9.36). Note that the argument of the Bessel function is the distance between a pair of feet multiplied by the parameter ρ. The parameter ρ captures the difference in mass density between the particles and the fluid.[5] The term c_{ijkl} originates in the coefficient of the force we examined in Chapter 5. Here, this coefficient is

$$c_{ijkl} = 2\pi\gamma Q_{ij}Qkl. \tag{9.37}$$

The terms Q_{ij} and Q_{kl} denote the wettability coefficient of the jth foot of the ith tile and the lth foot of the kth tile. Here, γ denotes the surface tension of the liquid on which the waterbugs float. It is also necessary to assume that $||w_{i,j} - w_{k,l}|| > 2R$, where R is the radius of a foot. This assumption ensures that feet are not touching. The total potential energy of the system can now be expressed as

$$U(q) = \sum_{1 \leq i \neq k \leq n} u(q_i, q_k). \tag{9.38}$$

We should note that there is an assumption here. We are assuming that the total potential energy can be obtained by simply summing over pairwise interactions of the particles. This requires that the disturbance of the meniscus between particles i and j due to other particles is negligible. This assumption may or may not be valid in a given system, yet it is a common simplifying assumption and serves to make the model tractable.

The kinetic energy is more easily obtained. The kinetic energy of a single tile is

$$K_i = \frac{m}{2}(\dot{x}_i^2 + \dot{y}_i^2 + d^2\dot{\theta}_i^2) \tag{9.39}$$

and hence the total kinetic energy is given by

$$K = \sum_{i=1}^{n} K_i. \tag{9.40}$$

Here, m is the mass of a waterbug and dots denote differentiation with respect to time.

Now, in the absence of friction, the equations of motion are obtained by defining the Lagrangian as $\mathcal{L} = K - U$ and applying the appropriate Euler-Lagrange equation. To include friction, the Lagrangian can be equated to the sum of the frictional forces and again the appropriate Euler-Lagrange equation

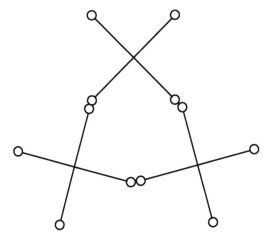

FIGURE 9.8:   An example of a defective waterbug assembly.

applied. Here, this yields

$$8m\ddot{x}_i + \sum_{k=1,k\neq i}^{2} \frac{\partial}{\partial x_i} u(q_i, q_k) = -4k_f \dot{x}_i \qquad (9.41)$$

$$8m\ddot{y}_i + \sum_{k=1,k\neq i}^{2} \frac{\partial}{\partial y_i} u(q_i, q_k) = -4k_f \dot{y}_i \qquad (9.42)$$

$$8md^2\ddot{\theta} + \sum_{k=1,k\neq i}^{2} \frac{\partial}{\partial \theta_i} u(q_i, q_k) = -4k_f d^2 \dot{x}_i. \qquad (9.43)$$

Analytically this model is intractable. To study the behavior, Klavins turned towards numerical simulation. But, one additional modelling assumption was necessary. As formulated, the model does not prevent or account for the contact of feet. In order to simulate this model numerically, Klavins assumed that when two feet made contact, they became the same foot. That is, they overlapped. Further, the singularity in the potential energy that occurs when feet made contact was smoothed for numerical purposes. We note that numerical simulations of this system are not trivial. Computations involving only 40 particles took several hours. Nevertheless, in numerical simulations, the waterbug model did appear to capture the appropriate self-assembly dynamics.

Klavins was able to analytically investigate the stability of assemblies of waterbugs that formed during his simulations. In particular, he was able to show that the assembly of two tiles in Figure 9.7 (a) was stable, while the assembly in Figure 9.7 (b) was not. Similar results could be obtained for systems of more than two tiles. But, the original intent of the model was not simply to investigate how waterbug tiles self-assembled, but to investigate how dynamic control over the wettability of the waterbug feet could translate into control over the self-assembling system. To study this, Klavins posed two problems. The first problem was to use the control over wettability to eliminate defects in a given self-assembled structure. An example of such a defect is shown in Figure 9.8. The second problem was to use the control over wettability to build finite structures. This is reminiscent of the problem posed by Hosokawa et al. who wanted to build stable structures consisting of only four particles. Klavins showed that standard ideas from *control theory* could be applied to this problem. In particular, he exhibited an open-looped controller that specified the wettability of individual feet as time evolved and could be used to eliminate a certain class of defects. Similarly, he exhibited a control scheme that allowed the construction of certain finite structures.

The Klavins waterbug model illustrates how a theoretical framework can help aid in the development of future self-assembling systems. In developing a relatively flexible model, Klavins was able to provide experimentalists with a platform to test and develop control schemes before costly experiments were conducted. It also illustrates some of the difficulties with theoretical approaches to self-assembly. In particular, even relatively simple models can quickly become analytically intractable and computationally challenging.

9.3 Abstract Models

In this section, we present three models that are abstracted from the general behavior of self-assembling systems. These models are representative examples of this approach. We refer the reader to [2, 3, 102, 123, 132] for information on similar abstract approaches to modelling self-assembly.

9.3.1 Conformational Switching

In Chapter 3, we learned that proteins can assume different *conformations*. For example, in examining the tobacco mosaic virus, we saw how interaction with RNA could induce washer shaped protein subunits to switch to a lock-washer conformation. In this new conformation, lock washers could bind together and ultimately form the complete viral unit. This is a typical example of a *conformational switch* in self-assembly; in one conformation a particle is unable to bind, but once it switches to a new state, binding becomes possible.

The first abstract model we'll consider is a model developed by Kazuhiro Saitou and Mark J. Jakiela [112, 113, 114] that attempts to clarify the *encoding power* of conformational switches. In particular, Saitou and Jakiela addressed the yield problem in self-assembly. They realized that a self-assembly process could be viewed as a sequence of small steps. This sequence, their *subassembly sequence*, dictated the final structures that were formed. Saitou and Jakiela wanted to know if particles capable of switching between different conformations could be used to encode for a particular subassembly sequence. In turn, if the subassembly sequence could be specified and unproductive sequences ruled out, they speculated that perhaps this would increase the yield of the final product.

First, to understand how a conformational switch might encode a subassembly sequence, let's consider a highly simplified situation. Imagine we have a system consisting of a mixture of three particle types. We'll label these A, B, and C. Let's picture our particles as squares and imagine that each particle has certain bond sites on its surface. We'll take our first set of particles to be as shown in Figure 9.9. Here, we'll use the same notation for bond sites

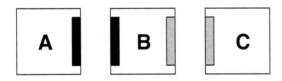

FIGURE 9.9: Particles and bonds for our first look at conformation switching.

that we did in discussing the DNA tiles of Chapter 8. In this system, that means that A can bond to B on the left and that C can bond to B on the right. Note that B cannot bond to itself, it can only bond to a matching site on a complementary particle. The same is true for A and C. Now, imagine that we have a large bag containing an equal mixture of these three tile types. At each time step, we'll reach into the bag, randomly grab two particles or clusters, pull them out, and if they can be bound, we'll bind them. We then return either the particles or the bound cluster to the bag and repeat the process. With the three particle types of Figure 9.9, it is clear that if we repeat this process long enough, we'll be left with a bag of ABC clusters. But, in constructing these clusters, there are two possible subassembly sequences. We can envision these as reactions. In particular,

$$A + B \longrightarrow AB \tag{9.44}$$
$$B + C \longrightarrow BC$$
$$AB + C \longrightarrow ABC$$
$$A + BC \longrightarrow ABC.$$

This means that we can get to the ABC clusters by either first forming a BC and then adding an A, or by first forming an AB and then adding a C. There is no preferred subassembly route.

To conveniently represent subassembly sequences, Saitou and Jakiela introduced the following notation. We write a complete subassembly sequence using parenthesis, at each step in the process placing a set of parenthesis around a completed subassembly. So, the subassembly sequences above would be denoted $(A(BC))$ and $((AB)C)$. Working our way inwards from the outermost pair of parenthesis, we recover the subassembly sequence in reverse. So, $(A(BC))$ denotes the sequence where BC formed first and an A was added, while $((AB)C)$ denotes the sequence where AB formed first and a C was added. A second way of viewing possible assembly sequences of a complete collection of particles was also introduced. In this notation, we form a tree as shown in Figure 9.10. Here, we begin in the top bin with a set of only

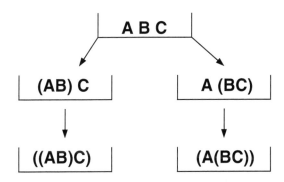

FIGURE 9.10:   Tree notation for representing assembly sequences.

three particles, A, B, and C. We then follow the arrows downwards through each possible assembly sequence. In either case we arrive at the end product, ABC.

Whatever our notation, it is clear that in the particle system of Figure 9.9 all subassembly sequences are equally likely. We can create a preferred route or preferred assembly sequence by introducing a conformational switch. In particular, we introduce a switch in the B particles. The type of switch we'll use was called a *minus device* by Saitou and Jakiela. A mechanical version of the switch is pictured in Figure 9.11. The binding sites on the new B particle are as before, but jutting out from the sides of the particle are push rods that can act to restrict binding. If a C particle attempts to bind with B the push rod will prevent this binding. On the other hand, if an A particle attempts to bind with B on the left, it will be able to push the push rod inwards and complete the bond. When this happens, the interior block slides

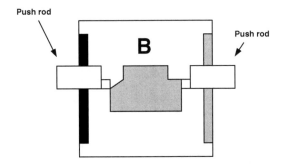

FIGURE 9.11: A mechanical conformational switch. The push rod on the left moves when an A particle attempts to bind. This pushes the center block downwards, freeing the right side for binding with a C particle.

downward, this frees the right hand push rod and makes the right hand bond site accessible. When an A binds with B and frees the right hand bond site, we say that B has switched its conformation. We denote this new conformation by B'. In our new system, the set of possible reactions is restricted. In particular, we have

$$A + B \longrightarrow AB' \tag{9.45}$$
$$AB' + C \longrightarrow AB'C.$$

This means that only one subassembly sequence is possible, the sequence denoted by $((AB)C)$. The introduction of a conformational switch has allowed us to *encode* for a particular subassembly sequence.

But, how does this affect the yield of the process? In our ABC examples it is not easy to see this effect. In fact, there are two different ways that the presence of a conformational switch can effect the yield of a particular self-assembly process. To understand the first way, consider the two systems shown in Figure 9.12. In either case, imagine that our goal is to assemble an ABC complex. In this case, that would result in a complete square. In Figure 9.12 (a), the bond sites for both the B and C particles are exposed. Hence, the following reactions can occur

$$A + B \longrightarrow AB \tag{9.46}$$
$$AB + C \longrightarrow ABC$$
$$A + C \longrightarrow AC.$$

Note that here, once an AC complex has formed, there is no pathway leading to the desired end product. That is, there are two assembly sequences, $((AB)C)$ and (AC). The first leads to the desired product, the second does not. In Figure 9.12, we have the same system, but the A particle now has a

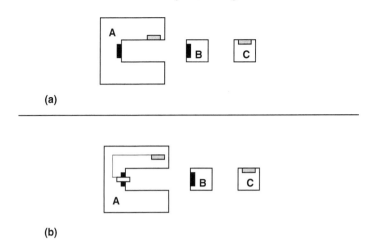

(a)

(b)

FIGURE 9.12: An example illustrating the effect of conformational switching on yield. In (a), the C particle can bind first, blocking the B particles. In (b), the bond site for C is unavailable until the AB bond has been made.

conformational switch. The bond site for C particles is not exposed unless a B particle binds first. The possible reactions for this system are

$$A + B \longrightarrow A'B \tag{9.47}$$
$$A'B + C \longrightarrow A'BC$$

and the only assembly sequence is $((A'B)C)$. The presence of the conformational switch prevents particles from getting trapped in the AC configuration and thereby increases the yield.

Saitou and Jakiela actually focused on the second way a conformational switch can effect they yield of an assembly process. To understand this effect, imagine we have a simple two particle system consisting of particle types A and B. This time, we imagine that A's can bond to B's from the left and hence the only reaction in the system is

$$A + B \longrightarrow AB. \tag{9.48}$$

We'll also imagine that our target complex is the AB complex. But, in this process, imagine that we have a large concentration of A's and a small concentration of B's. This implies that the number of favorable reactions occurring in the system will be very small. If we again imagine our particles being drawn from a large bag a pair at a time, most of the time we'll draw a pair of A's and return them to the bag with no bond being formed. This means that when stopped at some finite time, the yield of our process will be very low. Now, imagine we introduce a conformational switch. We'll let the A's bond with one another pairwise in the reaction

$$A + A \longrightarrow AA'. \tag{9.49}$$

The right hand A particle has switched to the conformation A'. This new conformation allows the A' to detach from the A when a bond with a B particle is formed. That is, the information about the bonding of B is propagated through the A' particle and delivered to the A particle. Saitou and Jakiela called this type of conformational switch a *plus device*. Here, this is represented by the reaction

$$AA' + B \longrightarrow A + AB. \tag{9.50}$$

We also imagine that once an AB is formed no further bonding is possible. Now, if we start with a high concentration of A's and a small concentration of B's, the concentration of A's will be rapidly reduced as AA' complexes form. This has the effect of increasing the frequency of reactions involving B and hence increasing the yield of the process.

Having seen the power of the conformational switch, let's consider a four particle system. We'll label our particles A, B, C, and D and allow the following reactions to occur

$$A + B \longrightarrow AB \tag{9.51}$$
$$B + C \longrightarrow BC$$
$$C + D \longrightarrow CD$$
$$AB + C \longrightarrow ABC$$
$$ABC + D \longrightarrow ABCD$$
$$B + CD \longrightarrow BCD$$
$$A + BCD \longrightarrow ABCD$$
$$AB + CD \longrightarrow ABCD.$$

Our target structure will be $ABCD$; there are five possible assembly sequences. They are: $(((AB)C)D)$, $((AB)(CD))$, $(A((BC)D))$, $(A(B(CD)))$, and $((A(BC))D)$. In the first system we considered, the ABC system, we could use the minus device to encode for any assembly sequence. However, in this four particle system, Saitou and Jakiela realized that the minus device was not powerful enough to encode for all of the different assembly sequences. In particular, they realized that $((A(BC))D)$ and $(A((BC)D))$ were *unencodable*, no matter how many conformational switches of the minus device type were used. Further, by simulating the set of reactions above numerically they observed that these unencodable sequences were among the highest yielding sequences in the system. In order to be able to encode these sequences using conformational switches, Saitou and Jakiela had to use both plus and minus devices. With this combination, any of the assembly sequences could be encoded.

But, now the general question remains. Suppose we have n labelled particles and we let them interact pairwise following the trend above. This quickly leads to enormous reaction sets and numerous assembly sequences. In examining the cases $n = 3$ and $n = 4$, Saitou and Jakiela had showed that their two conformational switch types were sufficient to encode for any assembly sequence. But what about the general case? Saitou and Jakiela focused on two major questions:

1. If we're given an assembly sequence can we tell whether or not it can be encoded using only minus devices? Or, some combination of minus and plus devices?

2. If a sequence can be encoded, how many conformations are necessary to encode the given assembly sequence?

Note that this second question was motivated by the four particle system. In order to encode the sequences $((A(BC))D)$ and $(A((BC)D))$ they had to use two types of switches. In addition, they had to use particles that carried both plus and minus device switches. These particles had more than two different conformations. Hence, if a combination of plus and minus devices was to be used, they had to allow for the possibility of particles with many conformations.

To answer these questions, Saitou and Jakiela defined a *one-dimensional self-assembling automaton* (SA). That is, they set up a formal structure, typical of those used in theoretical computer science, such that within the context of this structure they could answer the questions posed above. In their model, an (SA) was defined to be a pair $M = (\Sigma, R)$, where Σ was a finite set of components and R was a finite set of assembly rules. The assembly rules were of two forms. The first, was an attaching rule that abstracted the notion of minus devices. In particular, attaching rules were rules of the form

$$a^\alpha + b^\beta \rightarrow a^\gamma b^\delta. \tag{9.52}$$

Here, the a and b are particle labels, and the exponents track the conformation of the given particle. The second rule was a propagation rule abstracted from the notion of plus devices. These rules were of the form

$$a^\alpha b^\beta \rightarrow a^\gamma b^\delta. \tag{9.53}$$

Once this structure was in place, Saitou and Jakiela were able to classify SA's based on the presence of different rule types. They were able to prove that if both rule types were allowed any assembly sequence could be encoded using only three conformations per particle. The reader is referred to [112, 113, 114] for details of the proofs of these results.

The work of Saitou and Jakiela nicely illustrates the role of the abstract model in self-assembly. Their model was not based on any particular physical system, but rather on the basic notion of conformational switching seen in many self-assembling systems. With this type of model Saitou and Jakiela were able to ask and answer "What is possible?" type questions. Note that without this formal structure, little progress can be made with these questions. It is hard to imagine how within the context of one specific experimental system, questions such as these could be addressed. On the other hand, models such as this are always open to the criticism that they abstract too much. That too much of the real world is left out for the model to be able to be useful in designing any real world self-assembling system. The challenge is to bridge that gap, using the power of the abstract approach, but adding enough of the complexity of the real world to make such an approach useful in system design.

9.3.2 Graph Grammars

In Chapter 7, we discussed the dynamic self-assembling robotic tiles built by the Klavins' group. We noted that the tiles relied upon a *graph grammar* in order to make binding decisions. In this section, we'll examine this basic graph grammar approach as developed by Klavins and his collaborators [72, 73].

While the graph grammar approach incorporates the idea of conformational switching, it is aimed at different questions than the model of Saitou and Jakiela. To understand the questions addressed by the graph grammar approach, let's re-consider the simple self-assembling system we began with above. In particular, we'll again assume we have an equal mixture of three particle types, A, B, and C, and that they bind according to the reactions given in Equations (9.45). Now, above we assumed the complex ABC was our target structure. But, this time, let's imagine that we have two target structures, AB and CC. The first of these, AB, does occur during the assembly process. If we stopped the process at some finite time, we can expect to find some percentage of our particles in the AB state. But, the longer the process continues, the less of these particles we will find. Eventually, given enough time, all AB particles will bind with C's and form ABC complexes. Klavins et al. called structures like these *unstable*. This reflects the fact that these structures do not persist for all time. In contrast, structures like ABC are called *stable*. Once they form, no further change is possible. Our second target structure, CC, is an example of what Klavins et al. called *unreachable*. Given the particles and the binding interactions, there is no assembly process that will lead to the CC structure. Structures such as ABC, AB, or BC, are *reachable* structures within this system.

The graph grammar model is focused on these *stable* and *reachable* structures. Notice that the notion of a reachable structure is equivalent to the forward problem in self-assembly. The forward problem asks: Given a set of particles and a set of binding rules, what structures can form? In the language of the graph grammar model: Given a set of particles and a set of binding rules, what structures are *reachable*? A characterization of the set of reachable structures is one aspect of the graph grammar model. Klavins et al. also focused on the backward problem in self-assembly. In the language of the graph grammar model they asked: How can a set of binding rules be constructed to ensure that a desired target is reachable? To this they added the notion of *stable* structures. This is an aspect of the yield problem in self-assembly. By focusing on stable structures, Klavins et al. not only wanted to ensure that a target was reachable, but that target structures would appear once the assembly process was complete. In combining the questions about reachable and stable structures, Klavins et al. actually asked the very specific question: Is it possible to design a set of binding rules such that a desired target structure is the only stable element in the reachable set for the system?

To address this question, Klavins et al. worked within the context of the mathematical structure known as a *graph grammar*. A *graph* is simply a collection of vertices and edges. We encountered graphs in Chapter 6 when we studied the folding system of Griffith. If each vertex is given a name, we have a *labelled graph*. Formally, we say that a labelled graph, G, over an alphabet Σ, is a triple, $G = (V, E, l)$, where V is a set of vertices, E is a set of edges, and l is a labelling function. This function maps the vertices in V to the alphabet, Σ. That is, the labelling function gives each vertex a name. Throughout the remainder of this section, when we use the term *graph*, we'll mean a labelled graph.

Now, to build a graph grammar, we need to attach a rule set to our graph, G. The rule set simply describes what we usually think of as binding rules and conformational changes for a self-assembling system. As an easy example, let's develop a graph representation and a rule set for a simple system of particles. We imagine that we have one type of particle that can be in any one of three conformations. This means we need three letters in our alphabet, we'll use a, b, and c. To be concrete, let's assume that we start with eight particles all in conformation a. Each individual particle is identified by a vertex and labelled by a letter from our alphabet. Our initial setup is pictured in Figure 9.13. Note that initially our graph has no edges, only labelled vertices. The rule set captures how this graph evolves during an assembly process. As an example, consider the rule set, denoted Φ, defined by

$$
\begin{array}{lll}
a & a \longrightarrow b - b & \quad\quad (9.54) \\
a & b \longrightarrow b - c & \\
b & b \longrightarrow c - c. &
\end{array}
$$

The first rule says that a pair of vertices, each with label a, may be replaced by a pair of vertices, each with label b, and connected by an edge. The second and third rules are similar. These are examples of *constructive rules*. In the context of self-assembly, we are taking a pair of particles, creating a bond between them, and in the case of the rules above, changing their conformation. The action of this rule set on our initial graph is shown in Figure 9.13.

Notice that unlike the model of Saitou and Jakiela, here rules need not be applied one at a time. Rather, an assembly sequence is valid if some

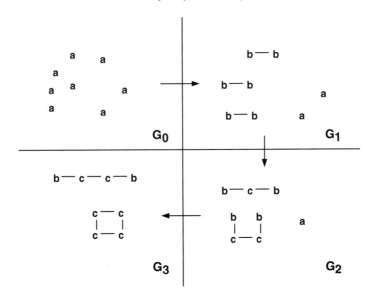

FIGURE 9.13: An example of an assembly sequence in the graph grammar model. Arrows indicate the direction of time.

application of the rule set to the graph at step i, G_i, produces the graph at step $i+1$, G_{i+1}. In Figure 9.13, we also see elements of the reachable set. Any structure that appears as we follow the arrows in the figure is, by definition, reachable. However, not all of these reachable structures are stable. We see that single particle pairings or chains of such particles are unstable. Only the closed loop of four c type particles is a stable structure. If you examine this structure and the rule set Φ, you'll see that no c's appear on the left hand side of any rules. Hence, no changes to this structure are possible and it must be stable.

Now, rules need not be constructive and they need not be limited to acting on only two vertices. We can have *destructive rules*, for example

$$b - b \longrightarrow a \quad a \tag{9.55}$$

or relabelling rules such as

$$b - b \longrightarrow a - c. \tag{9.56}$$

We can also have rules such as

$$c - c - c \longrightarrow c - a - c \tag{9.57}$$

that act on larger components of our graph. Putting the rules and graphs together, Klavins et al. defined their *graph assembly system* as the pair (G_0, Φ) consisting of an initial graph, G_0, and a rule set Φ.

Drawing on classical topological techniques, Klavins et al. were able to characterize the set of reachable structures and stable structures for their graph assembly system. More importantly, they were able to exhibit an algorithm that allowed them to construct a rule set Φ such that a given graph G constituted the entire reachable set for their graph assembly system. This argument also allowed them to find an algorithm that allowed them to construct a rule set Φ such that a given graph G constituted the entire stable set for their graph assembly system. Note that these algorithms are constructive. As their input they take a graph G and as their output they return a rule set Φ and a labelling function, l. The reader is referred to [72] for proofs of these results.

In some ways, the graph grammar approach of Klavins et al. is simply a generalization of the model of Saitou and Jakiela. However, it is an important generalization as it captures features of self-assembly not captured by the Saitou and Jakiela model. Additionally, it provides a constructive solution to an instance of the backward problem of self-assembly. Further, the graph grammar model has been implemented in the Klavins self-assembling robotic system. It serves as a language that allows the robotic particles to make decisions about binding and unbinding. Unfortunately, the model has its limitations. In particular, it does not account for geometry. Particles are treated abstractly as vertices with labels and there is no way to capture the geometric configuration of a given set of particles. Also, the algorithms developed by Klavins et al. can lead to rule sets that are not physically realizable. There is no way to prevent this within the context of the model. As with all such abstract models, the next step is to bring the model closer to physical reality.

9.3.3   The Tile Assembly Model

The *Tile Assembly Model* was introduced by Erik Winfree and developed by Winfree and his collaborators in [140, 141, 142, 107]. This important model links computation and self-assembly. As we saw in Chapter 8, by exploiting this link Winfree et al. were able to self-assemble a Sierpinski Gasket using DNA tiles. This link is the most promising route to programmed self-assembly available today.

To understand this connection between computation and self-assembly we need to introduce two concepts, the *Turing Machine* and *Wang Tiles*. The notion of a Turing Machine allows us to think about computation abstractly, i.e., divorced from any particular computer architecture or platform. The notion of Wang Tiles connects computation to arrays of tiles. The Tile Assembly Model marries the notion of Wang Tiles to the notion of self-assembly, thereby completing the connection to computation.

The Turing Machine was introduced by the mathematician Alan Turing in 1936 as a way of defining computation and as a tool for exploring the computability of objects. With this definition Turing formalized the question raised by David Hilbert in 1900, that is, whether it was possible or not to decide whether a given mathematical statement could be proved. In answering Hilbert's question, Turing also launched the field of theoretical computer science and provided its most powerful model.

A Turing Machine can be visualized as follows. Imagine we have an infinitely long tape that is subdivided into equal size squares. Each square can either be left blank, or can contain a zero or a one. Hovering over one square of the tape is a read-write head. This read-write head can erase a symbol, write a symbol, and advance the tape one square in either direction. The read-write head makes decisions about whether to read, write, or erase based on its internal state. The head has a finite number of states and a look-up table that dictates how it should behave once it reads the tape. That is, the head reads the value from the square below, checks its state, decides what its new state should be, what should be written in the square below, and how the tape should be advanced. The machine then executes this instruction and repeats. Usually we specify a special *halting state* where the Turing Machine stops all operation.

This definition of a Turing Machine can be made precise. It is typical to define a Turing Machine as a quintiple (S, A_T, N, s_0, F), where S is the finite set of available states, A_T is the tape alphabet, N is a transition function that tells the machine how to respond to a given input, s_0 is the initial state, and F is the set of halting states.

Whether we proceed formally or not, the important fact to remember about Turing Machines is that they provide a model of computation. Furthermore, this model is *universal*. Turing was able to show that there existed a *Universal Turing Machine* that could simulate any other Turing machine. This implies that any computing process that can be simulated by a Turing Machine can in fact be simulated by one machine, the Universal Turing Machine. One caveat - there are other possible models of computation. The famous *Church-Turing Thesis* lets us sidestep this complication. This thesis states that all sufficiently complex models of computation are equivalent. Hence, to study computation we need only study Turing Machines, and to study Turing Machines we need only study the Universal Turing Machine.

Now, suppose we have a model that we think is capable of simulating computation. In order to prove this, we simply need to prove that it is equivalent to a Universal Turing Machine, or more simply, we need to show that our model is *Turing Universal*. One such model was proposed by Hao Wang in 1961 [136]. In [136], he introduced what have become known as *Wang Tiles*. Wang tiles can be visualized as a set of square tiles colored or shaded as in Figure 9.14. For a given set of tiles, Wang imagined that infinitely many copies of each were available. Using these tiles, he then asked if they could be made to tile the plane. But, in placing each tile, one had to observe the rule that colored sides of adjacent tiles had to match. For example, the first tile in the figure could be placed below a copy of the second tile, but nowhere else. Tiles were not allowed to be rotated.

Profile - Erik Winfree

Erik Winfree is the architect of the *Tile Assembly Model*. With his introduction of this model in 1998, Winfree showed that not only was self-assembly equivalent to computation, but that implementing computation in self-assembly was practical. His method for constructing computing self-assembling tiles opened the door to the world of programmable self-assembly. This promises to usher in a new era of self-assembled, atomically precise, highly structured nanomaterials.

Fully embracing the interdisciplinary nature of self-assembly, Winfree has not only made seminal contributions to theoretical self-assembly, but to experimental self-assembly as well. In conjunction with Nadrian C. Seeman, Winfree demonstrated that DNA tiles could self-assemble into ordered periodic structures. A few years later, this time working with Paul W.K. Rothemund, Winfree fabricated a Sierpinski Gasket from DNA tiles, thereby demonstrating that the full promise of the Tile Assembly Model could be realized experimentally. For this work, Rothemund and Winfree shared the 2006 Foresight Institute's Feynman Prize for Nanotechnology – capturing the prize in both the experimental and theoretical categories.

In a recent conversation, Winfree shared his thoughts on the blind spots facing researchers in self-assembly today. He writes:

> *This is one blind spot: many researchers look at information and algorithms and can only see data processing programs running on conventional electronic computers – they don't see that information and algorithms are intrinsic to the behavior of molecular systems and that understanding this aspect of molecular behavior is key to fully exploiting what molecules can be designed to do. Or they do see that, but they don't yet see that it is no longer just the realm of science fiction.*

Winfree was trained as a mathematician and a computer scientist. He received an undergraduate degree in mathematics from the University of Chicago and a Ph.D. in Computation and Neural Systems from the California Institute of Technology. He is the recipient of numerous prizes and awards including a MacArthur Fellowship and a PECASE award from the National Science Foundation. In 1999, MIT's *Technology Review* named Winfree one of their "Top 100 Young Innovators."

Presently, Winfree is an associate professor of computer science at the California Institute of Technology. He currently leads several research efforts in the areas of DNA self-assembly, DNA computing, the study of gene regulatory networks, and DNA and RNA folding.

In his original paper, Wang conjectured that any set of tiles that could be placed so that they tiled the plane, did so periodically. In 1966, Robert Berger constructed a set of Wang tiles that tiled the plane, but aperiodically. At this time, it had already been shown that any Turing machine could be represented in terms of Wang tiles. By providing an example of an aperiodic Wang tiling, Berger showed that the tiling question of Wang was the same as the halting question in the theory of Turing Machines. This crucial result established that Wang tiles were in fact Turing Universal.

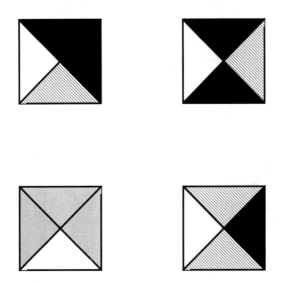

FIGURE 9.14: A set of four Wang tiles. Tiles must remain oriented as shown and can only be placed next to tiles if the edge colors match.

The crucial insight of Winfree was to realize that the colors on Wang tiles could be replaced by specific binding rules. But, binding rules meant self-assembly and that meant that tile based self-assembly could *compute*. In the other direction, this meant that tiles could be designed to produce programmed structures. Fully programmable self-assembly was possible.

Winfree did more than simply establish this result theoretically. With his Tile Assembly Model, Winfree showed how to construct tiles capable of computation via self-assembly that were physically realizable. The DNA tile systems of Chapter 8 are the most prominent examples of this construction. To specify the Tile Assembly Model, Winfree first needed to adapt Wang's notion of a tile. This was straightforward; a tile in the Tile Assembly Model *is* a Wang tile, but one described in the language of self-assembly. Instead of thinking of a tile as having colors, Winfree thought of his tiles in terms of *binding domains*. A tile in the Tile Assembly Model is a unit square with labelled edges and just like Wang tiles, these tiles cannot be rotated. The edges of these tiles possess bond sites on their four sides, and it is by matching bond sites that tiles can assemble.

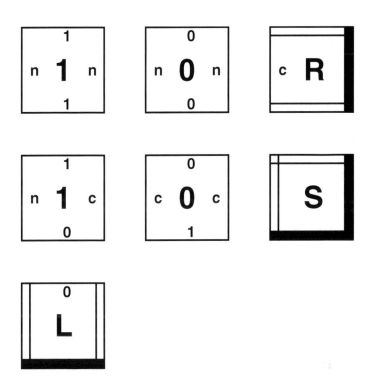

FIGURE 9.15:   Tiles in the Tile Assembly Model. Small characters denote binding sites, large characters specify tile type, and edge thickness indicates bond strength.

A simple set of seven tiles is shown in Figure 9.15. Note that each tile is labelled in three ways, a large character in the center, small characters around the edges, and by edge colors. The small characters represent the different types of binding sites. In this set, c's can bind to c's, 1's can bind to 1's and so forth. The edge colors indicate the *strength* of these possible bonds. The dark edges denote binding sites that have zero strength. The single lined edges denote binding sites that have strength one. The double lines indicate edges with strength two. The large characters in the center of each tile are used to denote tile type. The tiles with the L, R, and S are called *nucleating tiles*.

These tiles provide a nucleating center around which other tiles can grow and compute. The tiles with the 1 and the 0 are called *rule tiles*. These are the tiles that actually perform computation in the Tile Assembly Model.

Now, notice that in the Tile Assembly Model, tiles don't simply bond, they do so with a particular strength. By introducing this notion of bond strength, Winfree was also able to introduce the notion of *temperature* into the Tile Assembly Model. Suppose we denote the temperature of the system by τ. We can now restrict possible binding interactions by only allowing a tile to bind to the assembly if the sum of the bond strengths it makes with that assembly exceeds or equals the temperature, τ. For example, suppose we set $\tau = 1$ and we start with an L tile. Then, any tile that can match bond sites with the L tile can join the assembly. However, if we take $\tau = 2$, the growth is more restrictive. Initially the only tile that can join the L tile is an S tile; it's the only one capable of making a strength two bond. But, as the assembly grows we begin to encounter sites were a tile labelled 1 or 0 can make two bonds in joining the assembly. Since the sum of the bond strengths made by such tiles is now greater than or equal to the temperature, they are allowed to join. This type of bonding is called *cooperative bonding*. An example of a growing structure with temperature $\tau = 2$ is shown in Figure 9.16. This

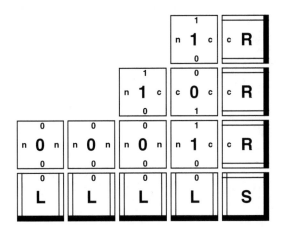

FIGURE 9.16: A sample configuration of tiles in the Tile Assembly Model. Here, the temperature, $\tau = 2$.

structure began with a single L tile, added an S tile to the right, added an R tile above the S, and then finally, a 1 could bind in the corner.

To relate this growing structure to computation, we interpret the rows in the structure, read from the bottom up, as output bit strings. Computation is performed on these bit strings by the tiles labelled 1 and 0 in the same way as computation was performed by the Rothemund system of Chapter 6. The lower and right edges of the 1 and 0 tiles serve as inputs. In matching bond sites on the growing structure, these edges are taking two tiles as input into a computation. The upper and left edges of the 1 and 0 tiles serve as the output. These output surfaces are exposed and available to serve as input for the next step of the computation.

Now, as with previous systems we've encountered, the Tile Assembly Model can be formally defined. In particular, a tile system, T, may be defined as a quadruple, (T, S, g, τ), where T is the set of tiles, S is a set of τ-stable *seed assemblies*, g is the strength function, and τ is the temperature. Here, the *strength function*, g, simply makes precise the notion of summing bond strengths, and the definition of τ-stable makes precise the notion of assemblies forming constrained by the temperature of the system. Within this model, self-assembly is a means of going from one configuration of a tile system to another configuration. A tile may be added to a configuration if the result is still τ-stable; that is, tiles must satisfy the temperature rules outlined above.

Within the context of the Tile Assembly Model, Winfree and his collaborators have been able to answer a wide range of questions. Here, we'll describe just one of these results. The reader is referred to [107, 140, 141, 142] for other related results.

In [107] Rothemund and Winfree sought to study the complexity of self-assembly. Recall that in Chapter 5, we introduced the notion of Kolmogorov complexity. We defined this measure of complexity for bit strings and loosely said that the complexity of a bit string was the length of the shortest computer program needed to produce that string. As an example, we considered strings of magnetic cubes created by self-assembly. If we started with cubes of two colors, one containing two north faces, the other two south faces, and labelled these by 1 and 0 respectively, we produced strings that looked like 01010101. That is, we produced an alternating colored string of cubes. If we started with eight cubes, four of each type, we would produce this result exactly. Since our assemblies were not oriented we treated this string and the string 10101010 as identical. In terms of the Kolmogorov program length definition, we could compute the complexity of this string. But, we can also relate this notion of complexity to our self-assembling system. The method of Winfree et al. is simply to count the number of distinct particle types needed to uniquely self-assemble a target object. That is, the complexity of a self-assembled object is the number of distinct particle types needed to guarantee that this object is the unique result of the self-assembly process. In this example, the complexity would be 2. We are able to assemble a string of alternating colored cubes using only two distinct particle types.

A less trivial question asks: What is the complexity of a self-assembled $N \times N$ square?[6] The step from one dimension to two dimensions is large. Notice that for our alternating colored cube example, the complexity of assembly was 2, independent of the length of the target object. It takes two particle types to assemble an alternating colored chain of length 8 or of length 800. This self-assembling system is not very complex. But, to guarantee that we assemble a square *and nothing else* requires a more complex system. If you think back to the many tile assembly systems of Chapter 6, you'll realize that none of them are capable of this task. Winfree et al. proved that the complexity of an $N \times N$ square was N^2 for systems with temperature $\tau = 1$. From the discussion of temperature above, we see that in the $\tau = 1$ case, bonding is not cooperative. A tile either bonds using one edge, or it cannot bond at all. When $\tau = 2$, Winfree et al. proved that the complexity dramatically decreased to $O(\log(N))$.[7] In this case, the bonding is cooperative. Tiles can bond either by activating a bond site of strength two, or by activating two bonds sites of strength one. That is, two particles already in the assembly could cooperate to add this new particle.

The Tile Assembly Model is an important abstract model of self-assembly. While it is abstracted from a wide range of self-assembling systems and while there are nonphysical aspects to the model, it does closely capture the behavior of self-assembling tile systems. In particular, it captures the behavior of DNA tiles; in fact, the design of DNA tiles was at least partially inspired by the Tile Assembly Model. Even an abstract model can be closely coupled to experiment. Further, the model allows us to address the question of the complexity of a self-assembling system. In addressing this question, the Tile Assembly Model provides useful answers to "What is possible?" type questions. Finally, it is the model that makes the important link between computation and self-assembly; this is likely the route that future progress will follow to achieve programmed self-assembly.

9.4 Chapter Highlights

- Theoretical approaches to self-assembly are widely varied. They are best defined by the questions they seek to answer.

- *Physical models* attempt to capture the behavior of one specific experimental system or some small set of closely related systems. They are useful for clarifying complicated parameter spaces and guiding experimental efforts.

- *Abstract models* attempt to answer "What is possible?" type questions. They are usually abstracted from a wide class of experimental systems. They are useful for clarifying minimal conditions under which a particular behavior of self-assembling systems can be achieved.

- The model of field driven mean curvature surfaces, the model of entropy driven helix formation, the chemical kinetics based model, and the waterbug model, are all examples of *physical models*. They all operate at different levels of abstraction, but are motivated by a small set of closely related experimental systems.

- The conformational switch model, the graph grammar model, and the Tile Assembly Model, are all examples of *abstract models*. They are motivated by a broad class of experimental systems or the general properties of self-assembly.

9.5   Exercises

Section 9.2

1. The parameter λ in Equations (9.15) and (9.16) contains the *square* of the applied voltage. Why? What does this imply for the possible shapes that can be obtained by applying an electric field?

2. Show that Equations (9.15) and (9.16) reduce to the catenoid problem of Chapter 6 when the applied voltage is removed.

3. Show that Equations (9.15) and (9.16) admit the special solution $u = 1$. Find the value of the parameters in the problem for which this solution exists.

4. Consider a hard sphere and rod model where two identical rods are in solution with a concentration of hard spheres. Suppose that the entropy and excluded volume ideas of this section hold, but that the rods are perfectly rigid and cannot bend. Instead, imagine that the rods are repulsive and that the energy needed to bring them together increases with the inverse of the square of the distance between their centers. Formulate a free energy and analyze this model. You may assume that the rods remain parallel and that their tops and bottoms are aligned.

5. The helix formation model of this section strongly resembles the magnetic system of Chapter 5 where an external field was applied. Discuss this relationship.

6. Return to the first two parts of this text and identify systems where an energy minimization principle came into play. Discuss how the modelling techniques of this section could be applied to construct models of those systems.

7. For the lattice model of this section, compute the possible configurations and energies for $N = 2$. Sketch a phase diagram for this system.

8. For the lattice model of this section, find two different rods, of the same length, that have the same total energy. Your rods should be truly different. That is, they should not be equivalent under rotation or reflection. Can you choose α and β such that your rods are the minimal energy configuration for your system?

9. The Hosokawa model is easily implemented on a computer. Construct a simulation using the parameter values $P_{11}^b = 0.438$, $P_{12}^b = 0.375$, $P_{22}^b = 0.25$, $P_{23}^b = 0$, $P_{13}^b = 0.188$, and $P_{33}^b = 0$. (Assume $P_{12}^b = P_{21}^b$ etc.) Plot x_1 versus x_2 for different instants in time. Do the same for x_3 versus x_4. Now, change P_{13}^b to zero and repeat. How do your results change?

10. The waterbug model is difficult to simulate numerically. However, based on the material of Chapter 5 and the ideas of this section, you should be able to build a similar, more tractable model. In particular, consider two interacting spherical particles on the surface of a fluid. In Chapter 5, we studied the attractive force between these particles. Use this force and the ideas of this section to build a dynamical model governing the attraction between two spherical particles. See [133] for one such model.

Section 9.3

11. Return to the self-assembling systems of the first two parts of this text and uncover all possible types of *conformational switching* that you can find. Compare and contrast these different types of switches. Which ones fit into the framework of the models of this section?

12. Numerically simulate the *ABC* model with and without a conformational switch. Keep track of the assembly sequences followed by the system. Compare the behavior of the two.

13. Again return to the first two parts of this book and find examples of self-assembling systems that can be described by the conformational switching model and examples of systems that cannot be described by this model. For one of those that can, formulate the model. For those that cannot, clearly explain what features of the system the model cannot capture.

14. Return to the first two parts of this book yet again and find examples of self-assembling systems that can and cannot be described by the graph grammar model. For one that can, formulate the initial graph and rule set. For one that cannot explain what features of the system the model cannot capture.

15. The Tile Assembly Model is closely related to DNA tiling as discussed in Chapter 8. Return to Chapter 8 and compare DNA tiling with the Tile

Assembly Model. What features of the physical system are captured? What features are left out?

16. The set of tiles presented here for the Tile Assembly Model are designed to count in binary. Extend the structure of Figure 9.16 and show how this binary counting takes place.

17. As noted in Chapter 5 and again here, the complexity of one-dimensional self-assembling systems are generally low. Devise a system that assembles a chain of length $3N$ such that the chain consists of repeating subunits that alternate colors, say, red, white, and blue. What is the complexity of your system?

18. Throughout this section, we avoided discussing defects in self-assembly. Each of the abstract models discussed incorporates or considers the possibility of defective bonding. How might the introduction of defects into these models affect the answers to the questions each model sets out to address?

9.6   Related Reading

The book by Lin and Segel is a great introduction to both mathematical modelling and continuum mechanics.

C.C. Lin and L.A. Segel, *Mathematics Applied to Deterministic Problems in the Natural Sciences*, SIAM, 1988.

If you wish to learn more about electromagnetism, Jackson and Stratton are the place to start.

J.D. Jackson, *Classical Electrodynamics*, Second Edition, Wiley, 1975.

J.A. Stratton, *Electromagnetic Theory*, McGraw-Hill, 1941.

An introduction to mathematical modelling in the context of micro- and nanosystems may be found in:

J.A. Pelesko and D.H. Bernstein, *Modeling MEMS and NEMS*, Chapman and Hall/CRC, 2002.

A good place to learn more about Lagrangian dynamics is in the book by Weinstock.

R. Weinstock, *Calculus of Variations*, Dover, 1974.

Abstract models of self-assembly often use the language and tools of computer science. You can learn more about these tool in the book by Sipser.

M. Sipser, *Introduction to the Theory of Computation*, PWS Publishing, 1997.

A gentler introduction to the ideas of theoretical computer science can be found in the books of Dewdney.

A.K. Dewdney, *The Tinkertoy Computer*, W.H. Freeman and Company, 1993.

A.K. Dewdney, *The New Turing Omnibus*, W.H. Freeman and Company, 1993.

9.7 Notes

1. The one exception to this is the first two abstract models, which are closely related and share notation.

2. If you've forgotten, the divergence theorem relates volume integrals to surface integrals. In this model, the integrals taken over the gap between the soap film and the electrode can be replaced by integrals over the soap film and electrode surfaces.

3. Note that this reference is to a Ph.D. thesis that was not yet complete when this book was written. I expect that it will be available by the time this book appears. If not, Derek Moulton will be happy to provide you with details of this work. He may be reached at moulton@math.udel.edu.

4. It is not yet clear whether or not this parameter regime is experimentally accessible *at all*.

5. In Chapter 5, ρ was q. Here q is used to denote the state of the system.

6. This problem was actually posed by Len Adleman. See profile in Chapter 2.

7. The "big O" notation is used in asymptotic analysis. The statement $O(\log(N))$ roughly means that as N tends to infinity the complexity grows like a constant times $\log(N)$.

Chapter 10

Directions

Imagination has this peculiarity that it produces the greatest things with as little time and trouble as little things.

Pascal, *Concerning the Vacuum*

10.1 Introduction

We began this journey with a goal – to understand the principles and techniques used by nature to self-assemble structures and to learn how to use these principles and techniques to design our own engineered self-assembling systems. In *Part I: The Natural World*, we saw examples of nature's technique and extracted key principles that reoccur throughout her design. In *Part II: Engineered Systems*, we saw how scientists and engineers use these principles and we learned a bit about the obstacles they face along the way. And, in this part of the book, *Part III: The Future*, we've surveyed theoretical approaches to self-assembly and learned how mathematical models are being used to push self-assembly technology forward. Along the way, we've examined dozens of natural and man-made self-assembling systems. We've chosen to highlight these systems for one of two reasons. Either, their simplicity helped us to easily understand the principles and problems of self-assembly, or they were landmark systems that have moved the science of self-assembly ahead significantly. But to paraphrase Hamlet, there are many more self-assembling systems under the sun than are written of in this book.[1] In this, the final chapter of *Self Assembly: The Science of Things that Put Themselves Together*, we briefly survey a collection of the most recent of these efforts. This time, we've chosen a mixture of examples to both illustrate broadly what is possible and examples that promise to quickly become technologically important. It's my hope, that having made it this far, you're ready to delve into the primary literature of self-assembly, grasp the essentials of each new development, and become an active participant in the field. The entry points into the literature provided in this chapter should help you figure out where to get started.

The first system we examine relates pineapples to self-assembly. The natural world often exhibits remarkable symmetries and patterns related to the integer sequence known as the *Fibonacci numbers*. In the system of Section 10.2, Li et al. [85] show that a stress-mediated self-assembly process in growing microspheres can lead to the same patterns. Our second system opens the door to technological advances in micro- and nanoelectromechanical systems (MEMS and NEMS). MEMS and NEMS technology has been hindered by the inherent planar nature of lithographic fabrication technology. In Section 10.3, we see that self-assembly may help MEMS and NEMS technology break free of the two dimensional world and start using three dimensional components. In Section 10.4, we return to biology, this time at the organism level. While nature clearly uses self-assembly to build from the cellular level on down, she also uses self-assembly to organize much larger organic systems. In a system consisting of thousands of sperm cells, we'll witness a self-assembly process strikingly similar to the dynamic self-assembling systems of Chapter 7. In Section 10.5, we'll look at even larger organisms, namely, humans. We'll see how the language and science of self-assembly is being used to describe and understand the emergence of social structures – from teams of Broadway producers to teams of scientists studying self-assembly. In the final example of this book, we return to where we began and examine the question of how biological molecules first made their appearance on earth. We see that self-assembly not only promises us a remarkable future, but may help us understand our own origins in the far distant past.

10.2   Fibonacci at the Nanoscale

A certain man had one pair of rabbits together in a certain enclosed place and one wishes to know how many are created from the pair in one year when it is the nature of them in a single month to bear another pair, and in the second month those born to bear also.

Fibonacci, *Liber Abaci, Book II*

With this brief passage, Fibonacci posed the question that would lead to the discovery of the integer sequence that bears his name, the famous *Fibonacci sequence*.[2] Now known to every student of mathematics, the Fibonacci sequence is defined by the recursion relation

$$x_{n+1} = x_n + x_{n-1} \tag{10.1}$$
$$x_0 = 1$$
$$x_1 = 1.$$

This sequence answers Fibonacci's rabbit question, but also shows up in a host of surprising places ranging from sunflowers to pineapples, and now can even be found embedded in the structure of self-assembled microspheres.[3]

In sunflowers, within the pattern of seeds in the head one can clearly identify two types of spirals. One type of spiral runs clockwise, the other counterclockwise. Counting the numbers of each type of spiral, we inevitably arrive at pairs, 21 and 34, or 34 and 55, or 55 and 89. These pairs are consecutive terms in the Fibonacci sequence. They arise because of the relationship among the Fibonacci sequence, the golden mean, and optimal packing. When objects such as seeds or thorns pack themselves on a spherical surface this sequence of spirals can also be observed. However, on spherical surfaces there is a second way of packing, namely hexagonal close packing, that also often arises. In these structures, the seeds or thorns arrange themselves like the bubbles in a bubble raft, but restricted to a spherical surface. Creating such an arrangement on a sphere necessarily gives rise to pentagonal or heptagonal defects. This is a consequence of the well-known Euler's rule for polyhedra.[4]

Recently, Chaorong Li, Xiaona Zhang, and Zexian Cao [85] showed that the

FIGURE 10.1: Natural and self-assembled structures illustrating triangular number patterns. (A) shows a silver core particle with a silicon oxide shell. This structure has a repeating hexagonal pattern. (B) shows a smaller, but similar particle. (C) is the same type of particle, but shows a pentagonal structure. (D) is a bud of the *Succisa pratensis Moench.* From Li, et al., Science, v. 309, pp. 909-911, (2005), Reprinted with permission from the AAAS.

naturally occurring patterns of sunflowers and pineapples could be produced on micron sized spheres of silver and silicon oxide. In this case, patterns were not produced as a result of a growth process, but rather were produced through a stress-mediated mechanism on quenched spheres exhibiting thermal mismatch. The Li et al. spheres had a silver center and were coated with a layer of silicon oxide. The spheres were grown through an evaporation process onto a sapphire substrate. The substrate was maintained at a temperature above the melting point of silver, but below that of silicon oxide. As the spheres were cooled the silicon oxide shell developed large stresses due to a mismatch in its thermal expansion coefficient with the silver core. If the cooling proceeds slowly, these stresses cause the surface to buckle, yet the coating remains attached to the core. As it buckles, it releases strain energy and develops patterns such as those shown in Figure 10.1. The small spherical bulges on the surface appear in a regular pattern. They exhibit either hexagonal close packing, or a sunflower-like sequence of spirals with Fibonacci sequence structure. Details may be found in [85].

10.3   Self-Assembly Springs Into Action

Micro- and nanoelectromechanical devices (MEMS and NEMS), are already all around us. The airbag in your car is triggered by a MEMS accelerometer. Your inkjet printer uses a piezoelectric MEMS device to produce uniform drops of ink. And, if you're on the cutting edge of home theater, your TV uses millions of tiny MEMS fabricated mirrors to create a high definition display. Yet, MEMS and NEMS technology faces one tremendous obstacle. The fabrication of MEMS and NEMS devices is based on the standard lithographic techniques used to fabricate integrated circuits. While there are tremendous advantages to using this well-characterized technique, it has the disadvantage of being inherently a two dimensional fabrication technology.

To take MEMS and NEMS into the third dimension requires the development of new fabrication techniques. One possible route is through self-assembly and in a recent effort Gao et al. [45] showed how to use self-assembly to fabricate nanoscale springs. This advance allows MEMS and NEMS researchers access to a common mechanical device, the spring, that was unobtainable via planar fabrication technology.

The springs of Gao et al. are true nanostructures with a diameter of about 300nm and a length of up to 100 microns. To fabricate these structures, Gao et al. used standard vapor-solid deposition techniques to grow zinc oxide crystalline strips. These crystals contain a superlattice long range structure consisting of a pair of crystals with their lattices oriented perpendicularly to one another. A small mismatch in this structure gives rise to a small strain

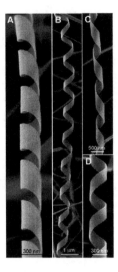

FIGURE 10.2: Self-assembled zinc oxide nanosprings. (A)-(C) show scanning electron microscope images of both left and right-handed coiled springs. (D) shows a close up illustrating the remarkable regularity and uniformity of the structures. From Gao, et al., Science, v. 309, pp. 1700-1704, (2005), Reprinted with permission from the AAAS.

in the flat growing strips. It is this strain that causes the strips to assume a regular helical formation.

The photographs in Figure 10.2 show the remarkable uniformity of the self-assembled nanosprings. Gao et al. extracted individual nanosprings from their batch and individually measured their elastic properties. For small displacements, these nanoscale springs behaved exactly like their macroscale counterparts. Details about the assembly process and characterization of materials properties may be found in [45].

10.4   Self-Assembled Swimming Cells

The biological systems we've examined in this book have all been at the sub-cellular level. Proteins exist inside of cells making up the cell's machinery, the ribosome is one example of this machinery, and the tobacco mosaic virus is small enough to hijack this machinery and make copies of itself. But, biological systems organize and self-assemble on scales larger than the cell. One example of this is the self-assembled vortex array of sea urchin spermatozoa recently uncovered by Riedel et al. [103].

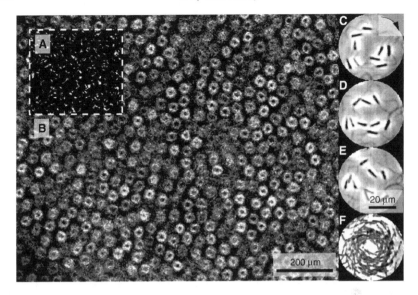

FIGURE 10.3: Self-assembled sea urchin spermatozoa. (A) shows the heads of the circulating sperm. (B) shows an averaged intensity over 25 frames of video. (C)-(E) show consecutive frames and (F) shows an average over 25 frames of a close up focusing on one vortex. From Riedel, et al., Science, v. 309, pp. 300-303, (2005), Reprinted with permission from the AAAS.

This system bears a striking resemblance to several of the dynamic self assembling systems we studied in Chapter 7. As you examine Figure 10.3 you should think back to the self-propelled particles of the Whitesides group and the mechanically driven particle system of Voth et al. In Figure 10.3, we see images at different scales of a collection of many thousands of sea urchin spermatozoa. Each individual sperm in this collection has a tail that propels it through the surrounding fluid. When isolated, this sperm will typically follow a helical path through space. But, when the surrounding fluid volume shrinks, so that the sperm's movement becomes confined to a thin fluid surface, this helical path switches to a circular path. When many thousands of such sperm are trapped on such a surface, the structure seen in Figure 10.3 emerges. Remarkably, this structure emerges due to hydrodynamic interactions; not chemical interactions. Even more remarkably, the sperm exhibit a phase transition dependent on sperm density just as with many of the inorganic systems of Chapter 7. Details of this system may be found in [103].

Profile - Paul Rothemund

Paul W. K. Rothemund has a nice smile. In fact, he has many of them. Of course, his smiles are only a few nanometers long and span a face only about one hundred nanometers across. And, he makes them himself in his laboratory.

Rothemund, a Senior Research Fellow at the California Institute of Technology, is one the pioneers of the nanoworld, teaching us how to self-assemble complicated structures from DNA. Rothemund works in the laboratory of Erik Winfree. Together, they have made seminal contributions to both theoretical and experimental nanoscience. In 2006, the pair shared awards in both the experimental and theoretical categories from the Foresight Nanotech Institute.

Rothemund's latest achievement is a method for constructing a DNA molecule that will self-assemble on a flat surface, taking on *any* desired shape. To illustrate his method Rothemund self-assembled the smiling faces shown below. In a recent conversation Rothemund commented on the future of self-assembly:

> *Our goal should be a programming language for molecules. Just as we have programming languages for computers, where we imagine a task a computer can perform, tell it what to do, and it does it, we should be able to do the same for molecules. Right now, DNA is the closest that we have but we should be able to do the same for other materials. While the problem of defining the rules that we'll need is wide open, we have some promising approaches. Algorithmic self-assembly is one of these.*

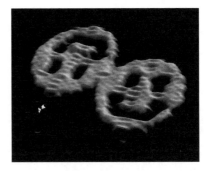

FIGURE 10.4: Paul W.K. Rothemund's self-assembled nanoscale smiling faces. These are atomic force microscope images. Credit: Paul W.K. Rothemund and Nick Papadakis.

10.5 Self-Assembly Goes Broadway

Moving still higher on the evolutionary tree, it has been found that even groups of humans self-assemble into large scale social structures. While this phenomenon has traditionally been the purview of those who study complexity and self-organization, it may be glimpsed through the lens of self-assembly; such a view may in fact help us to understand self-assembly more fully.

To convince yourself that structure can emerge among groups of humans without the need for governmental interference you should try the following simple experiment. Find a willing group of participants. The larger the better. Now, simply ask them to clap their hands and adjust their individual rhythms until the group is clapping in unison. You'll be shocked by how little time it takes.

At a higher level, humans also organize themselves into teams. In 2005, Guimera et al. studied how such teams self-assemble and how the process of self-assembly effects team performance [59]. Guimera et al. analyzed data characterizing collaboration in the Broadway music industry, social psychology, economics, ecology, and astronomy. Let's focus on the Broadway music industry so that we may understand the results of their study.

The group began by defining what was meant by a team in the industry. In the Broadway music industry this meant the collection of individuals who composed the music, wrote the lyrics and libretto, directed, produced, and designed the choreography. Those who acted in the production were excluded. From the data, they saw that the size of such teams had evolved with time. At the start of the industry, teams typically consisted of just two members. By the time the industry had matured, team size hovered around seven. But, they argued that team size was an insufficient description of team organization. They recognized that teams consisted of individuals embedded in a larger social network and that this network affected the formation of future teams.

Data on the network and on the characteristics of the individuals within the network was sparse, so the team turned to mathematical modelling to further their investigation. In their model, they built teams from individuals embedded in a growing network. Teams were built probabilistically and team formation was controlled by three parameters. These parameters were the number of team members, the probability of selecting an individual in the network to join the team, and the propensity of individuals to collaborate with those they've worked with in the past. They found that not only did the structure of the network effect the formation of teams, but that the team assembly mechanism effected the structure of the growing network. In fact, they observed a phase transition in their model. This transition occurred as the fraction of individuals in the network with past collaborations was increased from zero to one. At this transition point, the network shifted from a set of small disconnected clusters to one with a giant cluster containing most members of the network. For success on Broadway, it *is* who you know. Details of this study may be found in [59].

10.6   Self-Assembly and the Origin of Life

In the first paragraph of this book we observed - *No one put you together.* After ten chapters of studying self-assembly, perhaps this fact seems a little less remarkable. Hopefully, it seems a little more understandable. But, the question of the origin of life on earth, the question as to how the first cell arose, the question as to how the first biological molecules self-assembled, these questions still remain largely unanswered.

The most famous experimental answer to these questions was given by Stanley L. Miller and Harold C. Urey in 1953. In their system, Miller and Urey attempted to create environmental conditions resembling those present on the early Earth. They mixed a concoction of water, methane, ammonia, and hydrogen, placed it in a sealed system, added heat and electricity, and let the system percolate for seven days. At the end, they found that much of the carbon present at the start of their experiment had been transformed, had self-assembled, into amino acids. While the Miller-Urey experiment has been criticized, in particular, for not accurately capturing conditions present on early earth[5], the fact remains - they did demonstrate that biological molecules could self-assemble from inorganic compounds.

The quest to understand the origins of life on Earth, the conditions under which biological molecules can self-assemble, and the emergence of the original cell, continues. Unfortunately, in recent years much of the debate on these questions has become politicized. Fortunately, science marches onward. While we may not yet understand the origin of life, it is becoming clearer that we soon will, and that self-assembly will be a key.

10.7   Chapter Highlights

- Examples of self-assembling systems may be found in almost every scientific discipline. The systems of this chapter demonstrate the broad power of the concept of self-assembly.

10.8   Exercises

Section 10.2

1. Classify the example of this section as either static or dynamic self-assembly. What self-assembling systems from prior chapters does this system most closely resemble?

Section 10.3

2. Again, classify the example of this section as either static or dynamic self-assembly. How does it relate to the example of the last section? How does it relate to the model of helix formation (Chapter 9), magnetic origami (Chapter 5), and DNA Origami (Chapter 8)?

Section 10.4

3. Compare the self-assembling spermatozoa with the self-propelled particles of the Whitesides group discussed in Chapter 7. Recall that Whitesides was unable to perform experiments with large collections of these particles. If such an experiment were performed, what structures do you think might emerge? Would they resemble the spermatozoa structures? Or, would different structures occur?

Section 10.5

4. It's funny to think of humans as self-assembling particles, but that is the viewpoint taken by many of those who study emergent structures in social groups. In the clapping experiment and the team building system what types of particles are humans? That is, if we view them as structured particles, what features do they possess that are critical to the self-assembly process? How complicated would an artificial particle need to be in order to exhibit the synchronized clapping behavior of humans?

5. In the formation of teams in the Broadway music industry identify the four key components of self-assembly. What are the structured particles?

What is the binding force? What plays the role of the environment? What serves as the driving force?

10.9   Related Reading

There is an unbelievable volume of literature on the Fibonacci numbers. A good place to start is the web site of the Fibonacci Association.

The classic by Weyl explains many of the patterns found in nature.

H. Weyl, *Symmetry*, Princeton University Press, 1983.

If you are interested in MEMS and NEMS technology, there are several good introductions to the field. I recommend the book by Maluf as a good general introduction.

N. Maluf, *An Introduction to Microelectromechanical Systems Engineering*, Artech House, 1999.

The book by Madou gives a detailed introduction to fabrication techniques.

M.J. Madou, *Fundamentals of Microfabrication: The Science of Miniaturization, Second Edition*, CRC Press, 2002.

The collection by Trimmer contains all the classic articles on MEMS.

W. Trimmer (Editor), *Micromechanics and MEMS: Classic and Seminal Papers to 1990*, IEEE, 1997.

The book by Bonabeau, Dorigo, and Theraulz contains much information about self-assembling groups of organisms. Dorigo finds tremendous inspiration in ants.

E. Bonabeau, M. Dorigo, and G. Theraulaz, *Swarm Intelligence: From Natural to Artificial Systems*, Oxford University Press, 1999.

Other good examples of self-organization in humans may be found in the book by Colella, Klopfer, and Resnick.

V.S. Colella, E. Klopfer, and M. Resnick, *Adventures in Modeling: Exploring Complex, Dynamic Systems with StarLogo*, Teachers College Press, 2001.

If you are interested in the most up-to-date information about the origins of life question, I recommend starting with the Gordon Research Conference on the origins of life. A quick scan of the latest year's speakers will give you a good entry point into the literature.

Research advances in self-assembly appear in every type of journal. If you were to pick three to read regularly to keep abreast of advances in the field, I'd suggest *Science, Nature,* and the *Proceedings of the National Academies of Science.*

10.10   Notes

1. *There are more things in heaven and earth, Horatio, than are dreamt of in your philosophy.* Hamlet.

2. In the Liber Abaci, Fibonacci never writes down the sequence for which he would become famous. Neither does he ever construct its recursive definition. The Liber Abaci was in fact written by Fibonacci in order to introduce the western world to the Hindu number system.

3. There are many neat puzzles whose answer involves the Fibonacci numbers. My favorite is actually a tiling question. Given an $n \times 1$ strip of unit squares, in how many ways can this strip be tiled using dominos and unit squares? (Dominos cover two squares.)

4. Euler's rule says the $V - E + F = 2$, where V is the number of vertices, E is the number of edges, and F is the number of faces in the polyhedra.

5. Between 1953 and now, our picture of the early Earth has changed. This has led some to attempt to discredit the Miller-Urey experiment. To me, the question as to whether or not they got conditions exactly right seems to miss the point.

Chapter 11

Color Plates

HEY TUCK! WHAT ARE YA STARING AT?

SCIENTISTS CALL THAT SELF ASSEMBLY TUCK. IT'S HOW NATURE MAKES A LOT OF STUFF. AUTOMATICALLY THE UNCOATED CEREAL O'S ARE HYDROPHYLIC... MILK-O-PHYLIC IF YOU LIKE. THAT MEANS MILK LIKES TO COAT THEM.

THE UNCOATED O'S, SINK INTO THE MILK.

THE CHOCOLATE ONES ARE HYDROPHOBIC (MILK-O-PHOBIC) SO THE MILK DOESN'T COAT THEM.

MENISCUS

THEY FLOAT MORE ON THE TOP OF THE MILK.

WE CAN PROGRAM THE ASSEMBLY BY CHOCOLATE COATING EDGES OF THE UNCOATED ONES.

JUST DIP THREE SIDES INTO CHOCOLATE LIKE THIS!

NOW THEY TEND TO MAKE THESE FLOWER LIKE SHAPES.

KIND OF LIKE SNOWFLAKES!

SO IF I COAT THEM WITH JUST THE RIGHT PATTERNS THEY'LL MAKE ANY STRUCTURE I WANT?

MY CEREAL O'S ARE ASSEMBLING THEMSELVES INTO ROWS AND COLUMNS!

THE UNCOATED ONES ATTRACT EACH OTHER.

THE CHOCOLATE COATED ONES ATTRACT EACH OTHER.

AND THE CHOCOLATE COATED ONES REPEL THE UNCOATED ONES.

THEY SEEM TO BUNCH UP INTO GROUPS!

HEY TUCK YOURSELF HAS ASSEMBLED!

IT'S EVEN A MONSTER JUST LIKE YOU!

AHHHH!!

THAT WAS SURREAL... OR WAS IT CEREAL?

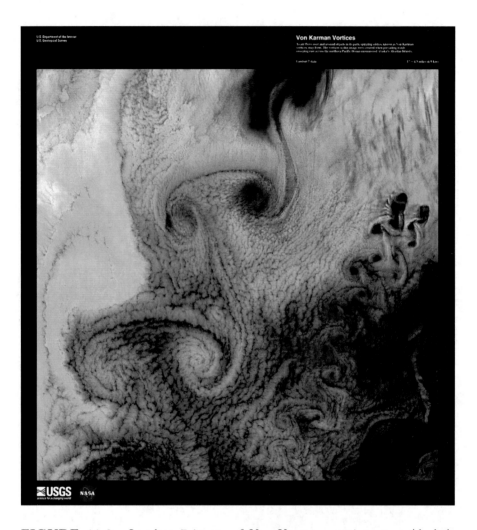

FIGURE 11.2: Landsat 7 image of Von Karman vortices over Alaska's Aleutian Islands. Image courtesy of USGS National Center for EROS and NASA Landsat Project Science Office.

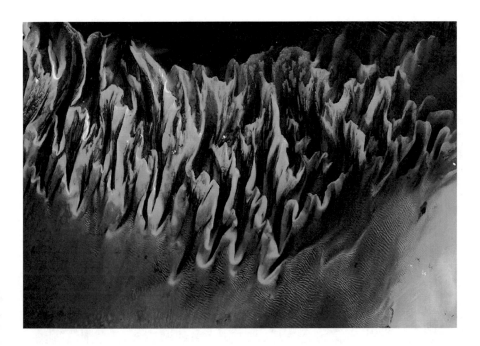

FIGURE 11.3:   Image of sand and seaweed in the Bahamas taken with the Enhanced Thematic Mapper plus (ETM+) instrument aboard the Landsat 7 satellite. Image courtesy of USGS National Center for EROS and NASA Landsat Project Science Office.

FIGURE 11.4: A bismuth "hopper crystal." Note the unusual lattice-work structure. Photograph by the author.

FIGURE 11.5: The Asian multicolored lady beetle, Harmonia axyridis, shows a unique pattern of spots. Credit: USDA Agricultural Research Service. Photo by Scott Bauer.

FIGURE 11.6:   Colored fluoride crystal. Credit: National Institutes of Health.

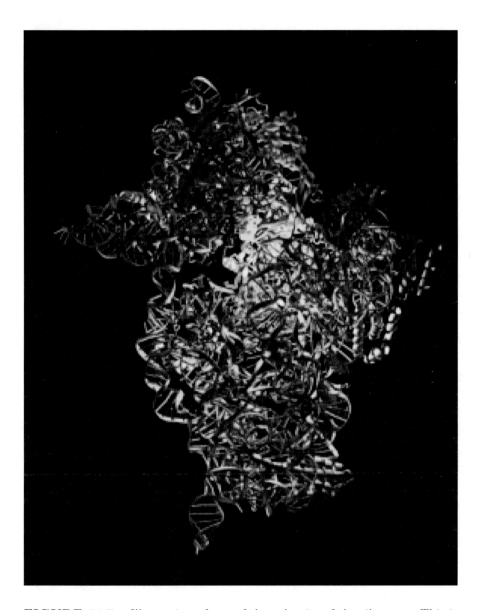

FIGURE 11.7: Illustration of one of the subunits of the ribosome. This is the "small" 30S subunit. Photo courtesy of the U.S. Department of Energy.

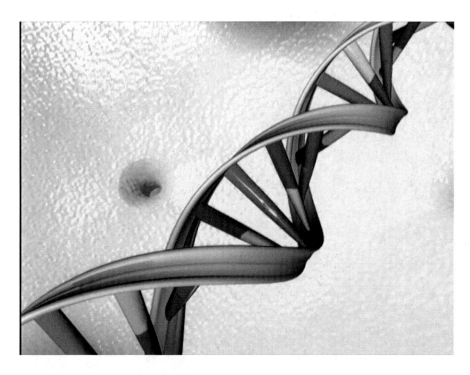

FIGURE 11.8: Illustration of the double helix structure of DNA. Credit: National Institutes of Health.

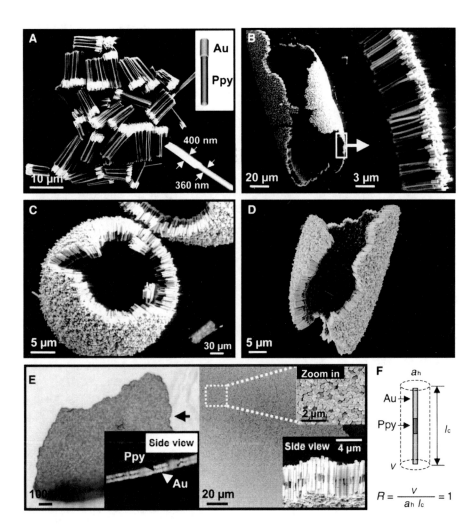

FIGURE 11.9: Images of self-assembled metal-polymer amphiphiles from the Mirkin group. From Park, et al., Science, v. 303, pp. 348-351, (2004), Reprinted with permission from the AAAS.

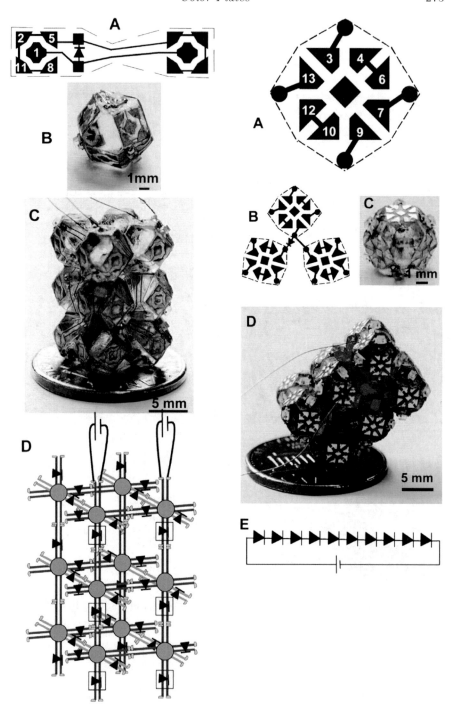

FIGURE 11.10: Images from the Whitesides group of three dimensional electrical networks constructed via self assembly. From Gracias, et al., Science, v. 289, pp. 1170-1172, (2000), Reprinted with permission from the AAAS.

FIGURE 11.11: Image of one of Eric Klavins' self-assembling programmable particles. Credit: Eric Klavins and Nils Napp.

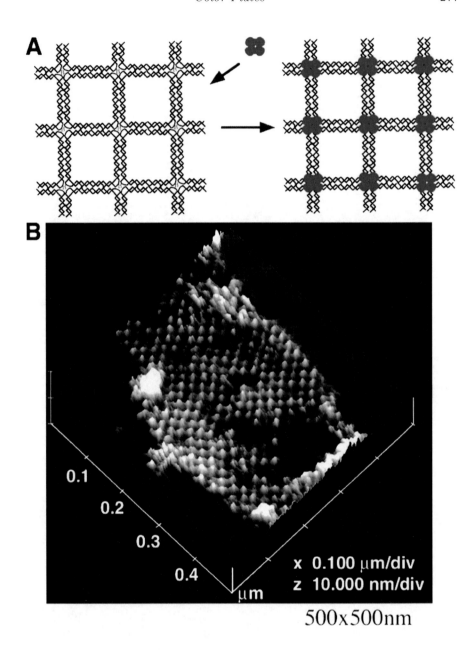

FIGURE 11.12: Atomic force microscope image of the Reif group's self-assembled protein array. Part (A) shows the tile units and selective binding of the protein molecule. Part (B) is the AFM image. From Yan, et al., Science, v. 301, pp. 1882-1884, (2003), Reprinted with permission from the AAAS.

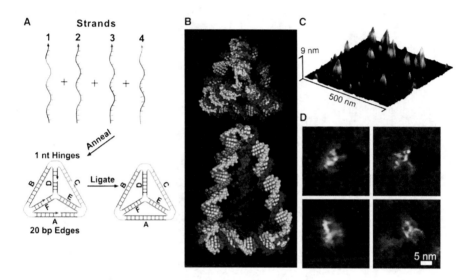

FIGURE 11.13:   Images of the self-assembled tetrahedra. (A) shows the assembly process. (B) shows the molecular structure of the tetrahedra. (C) shows atomic force microscope images of actual tetrahedra. From Goodman, et al., Science, v. 310, pp. 1661-1665, (2005), Reprinted with permission from the AAAS.

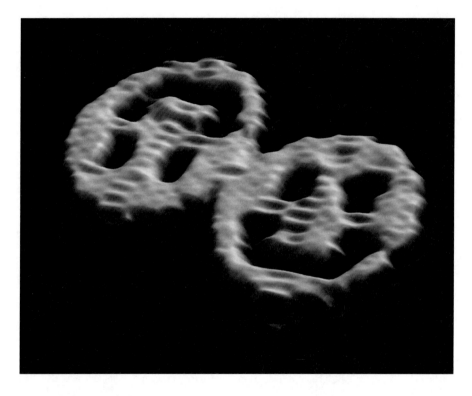

FIGURE 11.14: Atomic force microscope image of Paul Rothemund's DNA origami. Credit: Paul W.K. Rothemund and Nick Papadakis.

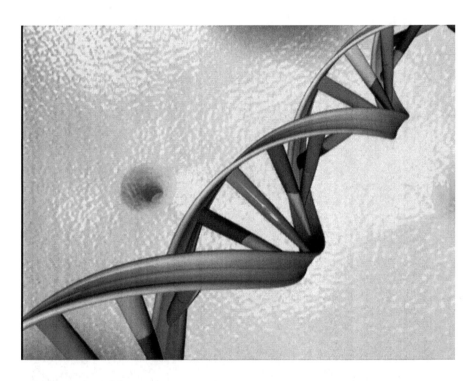

References

[1] L.M. Adleman, *Molecular Computation of Solutions to Combinatorial Problems*, Science, 266 (1994), pp. 1021-1024.

[2] L.M. Adleman, *Toward a Mathematical Theory of Self-Assembly*, Technical Report 00-722, Department of Computer Science, University of Southern California, (2000).

[3] G. Aggarwal, Q. Cheng, M.H. Goldwasser, M. Kao, P. Espanes, and R.T. Schweller, *Complexities for Generalized Models of Self-Assembly*, SIAM J. Comput., 34 (2005), pp. 1493-1515.

[4] B. Alberts, D. Bray, J. Lewis, M. Raff, K. Roberts, and J.D. Watson, *Molecular Biology of the Cell*, (1994), Garland Publishing, New York and London.

[5] C.B. Anfinsen, *Principles that Govern the Folding of Protein Chains*, Science, 181 (1973), pp. 223-230.

[6] I.S. Aranson, D. Blair, V.A. Kalatsky, G.W. Crabtree, W.K. Kwok, V.M. Vinokur, and U. Welp, *Electrostatically Driven Granular Media: Phase Transitions and Coarsening*, Phys. Rev. Lett. 84 (2000), pp. 3306-3309.

[7] I.S. Aranson, B. Meerson, P.V. Sasorov, and V.M. Vinokur, *Phase Separation and Coarsening in Electrostatically Driven Granular Media*, Phys. Rev. Lett. 88 (2002), 204301.

[8] T. Aste and D. Weaire, *The Pursuit of Perfect Packing*, (2000), Institute of Physics Press, Bristol and Philadelphia.

[9] P.A. Bachmann, P.L. Luisi, and J. Lang, *Auto Catalytic Self-Replicating Micelles as Models for Prebiotic Structures*, Nature, 357 (1992), pp. 57.

[10] P. Ball, *Algorithmic Crystals Made to Order*, Nature, 433 (2005), pp. 210.

[11] E.A. Bender, *An Introduction to Mathematical Modeling*, (1978), Dover, New York.

[12] B. Berger, P.W. Shor, L. Tucker-Kellogg, and J. King, *Local Rule-Based Theory of Virus Shell Assembly*, PNAS, 91 (1994), pp. 7732-7736.

[13] A. Bezryadin, R.M. Westervelt, and M. Tinkham, *Self-Assembled Chains of Graphitized Carbon Nanoparticles*, App. Phys. Lett. 74 (1999), pp. 2699-2701.

[14] J. Bishop, S. Burden, E. Klavins, R. Kreisberg, W. Malone, N. Napp, and T. Nguyen, *Programmable Parts: A Demonstration of the Grammatical Approach to Self-Organization*, Int. Conf. on Intelligent Robots and Sys. IEEE and RSJ Automation Society, (2005).

[15] M. Boncheva, S.A. Andreev, L. Mahadevan, A. Winkleman, D.R. Reichman, M.G. Prentiss, S. Whitesides, and G.M. Whitesides, *Magnetic Self-Assembly of Three-Dimesnional Surfaces from Planar Sheets*, PNAS, 102 (2005), pp. 3924-3929.

[16] M. Boncheva, D.H. Gracias, H.O. Jacobs, and G.M. Whitesides, *Biomimetic Self-Assembly of a Functional Asymmetrical Electronic Device*, PNAS, 99 (2002), pp. 4937-4940.

[17] N. Bowden, A. Terfort, J. Carbeck, and G.M. Whitesides, *Self-Assembly of Mesoscale Objects into Ordered Two-Dimensional Arrays*, Science, 276 (1997), pp. 233-235.

[18] N. Bowden, S.R.J. Oliver, and G.M. Whitesides, *Mesoscale Self-Assembly: Capillary Bonds and Negative Menisci*, J. Phys. Chem. B, 104 (2000), pp. 2714-2724.

[19] C.V. Boys, *Soap Bubbles: Their Colors and Forces Which Mold Them*, (1959), Dover, New York.

[20] L. Bragg, *A Model Illustrating Intercrystalline Boundaries and Plastic Flow in Metals*, J. Sci. Inst. 19 (1942), pp. 148-150.

[21] L. Bragg and J.F. Nye, *A Dynamical Model of Crystal Structure*, Proc. Roy. Soc. London, 190 (1947), pp. 474-481.

[22] T.L. Breen, J. Tien, S.R.J. Oliver, T. Hadzic, and G.M. Whitesides, *Design and Self-Assembly of Open, Regular, 3D Mesostructures*, Science, 284 (1999), pp. 948-951.

[23] D.A. Bruzewicz, M. Bonceva, A. Winkleman, J.M. St. Clair, G.S. Engel, and G.M. Whitesides, *Biomimetic Fabrication of 3D Structures by Spontaneous Folding of Tapes*, J. Am. Chem. Soc., 128 (2006), pp. 9314-9315.

[24] P.J.G. Butler and A. Klug, *The Assembly of a Virus*, Sci. Am., 242 (1978), pp. 62-69.

[25] D.J. Campbell, E.R. Freidinger, J.M. Hastings, and M.K. Querns, *Spontaneous Assembly of Soda Straws*, J. Chem. Ed. 79 (2002), pp. 201-202.

[26] J. Chen and N.R. Seeman, *Synthesis from DNA of a Molecule with the Connectivity of a Cube*, Nature, 350 (1991), pp. 631-633.

[27] J. Chen, N.R. Kallenbach, and N.C. Seeman, *A Specific Quadrilateral Synthesized from DNA Branched Junctions*, J. Am. Chem. Soc. 111 (1989), pp. 6402-6407.

[28] T.D. Clark, R. Ferrigno, J. Tien, K.E. Paul, and G.M. Whitesides, *Template-Directed Self-Assembly of 10-μm-Sized Hexagonal Plates*, J. Am. Chem. Soc. 124 (2002), pp. 5419-5426.

[29] A. Czirok, H.E. Stanley, and T. Vicsek, *Spontaneously Ordered Motion of Self-Propelled Particles*, J. Phys. A, 30 (1997), pp. 1375-1385.

[30] D. Deamer, S. Singaram, S. Rajamani, V. Kompanichencko, and S. Guggenheim, *Self-Assembly Processes in the Prebiotic Environment*, Phil. Trans. Roy. Soc. B, 361 (2006), pp. 1809-1818.

[31] K. DeBell, A.B. MacIsaac, and J.P. Whitehead, *Dipolar Effects in Magnetic Thin Films and Quasi-Two-Dimensional Systems*, Rev. Mod. Phys. 72 (2000), pp. 225-257.

[32] M. Doi, *Introduction to Polymer Physics*, (2001), Oxford Science Publications, Oxford.

[33] A. Docoslis and P. Alexandridis, *One-, Two-, and Three-Dimensional Organization of Colloidal Particles Using Nonuniform Alternating Current Fields*, Electrophoresis, 23 (2002), pp. 2174-2183.

[34] K.E. Drexler, *Engines of Creation, The Coming Era of Nanotechnology*, (1990), Fourth Estate, London.

[35] M. Dueweke, U. Dierker, and A. Hubler, *Self-Assembling Electrical Connections Based on the Principle of Minimum Resistance*, Phys. Rev. E, 54 (1996), pp. 496-506.

[36] K.E. Dungey, G. Lisensky, and S.M. Condren, *Kixium Monolayers: A Simple Alternative to the Bubble Raft Model for Close-Packed Spheres*, J. Chem. Ed. 77 (2000), pp. 618-619.

[37] D.M. Eigler and E.K. Schweizer, *Positioning Single Atoms with a Scanning Tunnelling Microscope*, Nature, 344 (1990), pp. 524-526.

[38] A. Ekani-Nkodo, A. Kumar, and D. Kuchnir Fygenson, *Joining and Scission in the Self-Assembly of Nanotubes from DNA*, Phys. Rev. Lett. 93 (2004), pp. 268301.

[39] P.F.N. Faisca and R.C. Ball, *Thermodynamic Control and Dynamical Regimes in Protein Folding*, J. Chem. Phys. 116 (2002), pp. 7231-7237.

[40] P.F.N. Faisca and R.C. Ball, *Topological Complexity, Contact Order, and Protein Folding Rates*, J. Chem. Phys. 117 (2002), pp. 8587-8591.

[41] M. Fialkowski, K.J.M. Bishop, R. Klajn, S.K. Smoukov, C.J. Campbell, and B.A. Grzybowski, *Principles and Implementations of Dissipative (Dynamic) Self-Assembly*, J. Phys. Chem. B, 110 (2006), pp. 2482-2496.

[42] R.P. Feynman, R.B. Leighton, and M. Sands, *The Feynman Lectures on Physics: The Definitive and Extended Edition*, (2005), Addison Wesley, New York.

[43] H. Fraenkel-Conrat and R.C. Williams, *Reconstitution of Active Tobacco Mosaic Virus from its Inactive Protein and Nucleic Acid Components*, PNAS, (1955), pp. 690-698.

[44] R.A. Freitas and R.C. Merkle, *Kinematic Self-Replicating Machines*, (2004), Landes Bioscience, Texas.

[45] P.X. Gao, Y. Ding, W. Mai, W.L. Hughes, C. Lao, and Z.L. Wang, *Conversion of Zinc Oxide Nanobelts into Superlattice Structured Nanohelices*, Science, 309 (2005), pp. 1700-1704.

[46] H. Gau, S. Herminghaus, P. Lenz, and R. Lipowsky, *Liquid Morphologies on Structured Surfaces: From Microchannels to Microchips*, Science, 283 (1999), pp. 46-49.

[47] A. Ghazali and J.C. Levy, *Two-Dimensional Arrangements of Magnetic Nanoparticles*, Phys. Rev. B, 67 (2003), 064409.

[48] J. Gleick, *Genius: The Life and Science of Richard Feynman*, (1992), Pantheon Books, New York.

[49] N. Go, *Theoretical Studies of Protein Folding*, Ann. Rev. Biophys. Bioeng., 12 (1983), pp. 183-210.

[50] M. Golozovsky, Y. Saado, and D. Davidov, *Self-Assembly of Floating Magnetic Particles into Ordered Strutures: A Promising Route for the Fabrication of Tunable Photonic Band Gap Materials*, App. Phys. Lett. 75 (1999), pp. 4168-4170.

[51] R.P. Goodman, I.A.T. Schapp, C.F. Tardin, C.M. Erben, R.M. Berry, C.F. Schmidt, and A.J. Tuberfield, *Rapid Chiral Assembly of Rigid DNA Building Blocks for Molecular Nanofabrication*, Science, 310 (2005), pp. 1661-1665.

[52] S. Govindarajan and R.A. Goldstein, *On the Thermodynamic Hypothesis of Protein Folding*, PNAS, 95 (1998), pp. 5545-5549.

[53] D.H. Gracias, J. Tien, T.L. Breen, C. Hsu, and G.M. Whitesides, *Forming Electrical Networks in Three Dimensions by Self-Assembly*, Science, 289 (2000), pp. 1170-1172.

[54] S.T. Griffith, *Growing Machines*, Ph.D. Thesis, Massachusetts Institute of Technology, (2004).

[55] B.A. Grzybowski, H.A. Stone, and G.M. Whitesides, *Dynamic Self-Assembly of Magnetized, Millimeter-Sized Objects Rotating at a Liquid-Air Interface*, Nature, 405 (2000), pp. 1033-1036.

[56] B.A. Grzybowski, X. Jiang, H.A. Stone, and G.M. Whitesides, *Dynamic, Self-Assembled Aggregates of Magnetized, Millimeter-Sized Objects Rotating at the Liquid-Air Interface: Macroscopic, Two-Dimensional Classical Artificial Atoms and Molecules*, Phys. Rev. E, 64 (2001), 011603.

[57] B.A. Grzybowski, H.A. Stone, and G.M. Whitesides, *Dynamics of Self Assembly of Magnetized Disks Rotating at the Liquid-Air Interface*, PNAS, 99 (2002), pp. 4147-4151.

[58] B.A. Grzybowski, M. Radkowski, C.J. Campbell, J. Ng Lee, and G.M. Whitesides, *Self-Assembling Fluidic Machines*, App. Phys. Lett. 84 (2004), pp. 1798-1800.

[59] R. Guimera, B. Uzzi, J. Spiro, and L.A. Nunes Amaral, *Team Assembly Mechanisms Determine Collaboration Network Structure and Team Performance*, Science, 308 (2005), pp. 697-702.

[60] H. Helgesen, A.T. Skjeltorp, P.M. Mors, R. Botet, and R. Jullien, *Aggregation of Magnetic Microspheres: Experiments and Simulations*, Phys. Rev. Lett. 61 (1988), pp. 1736-1739.

[61] K.D. Hermanson, S.O. Lumsdon, J.P. Williams, E.W. Kaler, and O.D. Velev, *Dielectrophoretic Assembly of Electrically Functional Microwires from Nanoparticles Suspensions*, Science, 294 (2001), pp. 1082-1086.

[62] K. Hosokawa, I. Shimoyama, and H. Miura, *Two-Dimensional Micro-Self-Assembly Using the Surface Tension of Water*, Sensors and Actuators A, 57 (1996), pp. 117-125.

[63] W. Huck, J. Tien, and G.M. Whitesides, *Three-Dimensional Mesoscale Self-Assembly*, J. Am. Chem. Soc. 120 (1998), pp. 8267-8268.

[64] C.R. Iacovella, M.A. Horsch, Z. Zhang, and S.C. Glotzer, *Phase Diagrams of Self-Assembled Mono-Tethered Nanospheres from Molecular Simulation and Comparison to Surfactants*, Langmuir, 21 (2005), pp. 9488-9494.

[65] R.F. Ismagilov, A. Schwartz, N. Bowden, and G.M. Whitesides, *Autonomous Movement and Self-Assembly*, Angew. Chem. Int. Ed. 41 (2002), pp. 652-654.

[66] J. Israelachvili, *Intermolecular and Surface Forces*, (1992), Academic Press, New York.

[67] H.O. Jacobs, A.R. Tao, A. Schwartz, D.H. Gracias, and G.M. Whitesides, *Fabrication of a Cylindrical Display by Patterned Assembly*, Science, 296 (2002), pp. 323-325.

[68] S.L. Keller, *Sequential Folding of a Rigid Wire into Three-dimensional Structures*, Am. J. Phys., 72 (2004), pp. 599-603.

[69] K. Keren, R.S. Berman, E. Buchstab, U. Sivan, and E. Braun, *DNA-Templated Carbon Nanotube Field-Effect Transistor*, Science, 302 (2003), pp. 1380-1382.

[70] E. Kim and G.M. Whitesides, *Use of Minimal Free Energy and Self-Assembly to Form Shapes*, Chem. Mater. 7 (1995), pp. 1257-1264.

[71] E. Klavins, *Toward the Control of Self-Assembling Systems*, Control Problems in Robotics, (2003), Springer, pp. 153-168.

[72] E. Klavins, R. Ghrist, and D. Lipsky, *Graph Grammars for Self Assembling Robotic Systems*, Proc. Int. Conf. Robotics and Automation, (2004).

[73] E. Klavins, S. Burden, and N. Napp, *Optimal Rules for Programmed Stochastic Self-Assembly*, Robotics: Science and Systems, Philadelphia, PA, (2006).

[74] A.N. Kolmogorov, *Three Approaches to the Definition of the Concept of "Quantity of Information,"* Probl. Inform. Transmission, 1 (1965), pp. 1.

[75] B.A. Korgel, *Self-Assembled Nanocoils*, Science, 303 (2004), pp. 1308-1309.

[76] X.Y. Kong, Y. Ding, R. Yang, and Z.L. Wang, *Single-Crystal Nanorings Formed by Epitaxial Self-Coiling of Polar Nanobelts*, Science, 303 (2004), pp. 1348-1351.

[77] P.A. Kralchevsky and K. Nagayama, *Capillary Interactions Between Particles Bound to Interfaces, Liquid Films and Biomembranes*, Adv. Coll. Int. Sci., 85 (2000), pp. 145-192.

[78] T.H. LaBean, H. Yan, J. Kopatsch, F. Liu, E. Winfree, J.H. Reif, and N.C. Seeman, *Construction, Analysis, Ligation, and Self-Assembly of DNA Triple Crossover Complexes*, J. Am. Chem. Soc. 122 (2000), pp. 1848-1860.

[79] R.G. Larson, *The Structure and Rheology of Complex Fluids*, (1999), Oxford University Press, Oxford.

[80] J.A. Lake, *The Ribosome*, Sci. Am., 245 (1981), pp. 84-97.

[81] P. Lenz, W. Fenzl, and R. Lipowsky, *Wetting of Ring-Shaped Surface Domains*, Europhys. Lett. 53 (2001), pp. 618-624.

[82] P. Lenz and R. Lipowsky, *Morphological Transitions of Wetting Layers on Structured Surfaces*, Phys. Rev. Lett. 80 (1998), pp. 1920-1923.

[83] P.E. Leopold, M. Montal, and J.N. Onuchic, *Protein Folding Funnels - A Kinetic Approach to the Sequence Structure Relationship*, PNAS, 89 (1992), pp. 8721-8725.

[84] C. Levinthal, *Are There Pathways for Protein Folding?*, J. Chim. Phys. 65 (1968), pp. 44.

[85] C. Li, X. Zhang, and Z. Cao, *Triangular and Fibonacci Number Patterns Driven by Stress on Core/Shell Microstructures*, Science, 309 (2005), pp. 909-911.

[86] J.D. Lohn, G.L. Haith, and S.P. Colombano, *Two Electromechanical Self-Assembling Systems*, Sixth Foresight Conference on Molecular Nanotechnology, (1998).

[87] G.P. Lopez, H.A. Biebuyck, C.D. Frisbie, and G.M. Whitesides, *Imaging of Features on Surfaces by Condensation Figures*, Science, 260 (1993), pp. 647-649.

[88] V.N. Manoharan, M.T. Elsesser, and D.J. Pine, *Dense Packing and Symmetry in Small Clusters of Microspheres*, Science, 301 (2003), pp. 483-487.

[89] E.H. Mansfield, H.R. Sepangi, and E.A. Eastwood, *Equilibrium and Mutual Attraction or Repulsion of Objects Supported by Surface Tension*, Phil. Trans. R. Soc. Lond. A, 355 (1997), pp. 869-919.

[90] C. Mao, W. Sun, and N.C. Seeman, *Assembly of Borromean Rings from DNA*, Nature, 386 (1997), pp. 137-138.

[91] C. Mao, V.R. Thalladi, D.B. Wolfe, S. Whitesides, and G.M. Whitesides, *Dissections: Self-Assembled Aggregates That Spontaneously Reconfigure Their Structure When Their Environment Changes*, J. Am. Chem. Soc., 124 (2002), pp. 14508-14509.

[92] F.S. Merkt, R.D. Deegan, D.I. Goodman, E.C. Rericha, and H.L. Swinney, *Persistent Holes in a Fluid*, Phys. Rev. Lett. 92 (2004), 184501.

[93] D. Moulton, *Mathematical Modelling of Field Driven Mean Curvature Surfaces*, Ph.D. Thesis, University of Delaware, to appear.

[94] F. Nadal, F. Argoul, P. Hanusse, and B. Pouligny, *Electrically Induced Interactions Between Colloidal Particles in the Vicinity of a Conducting Plane*, Phys. Rev. E, 65 (2002), 061409.

[95] R. Nagarajan and E. Ruckenstein, *Theory of Surfactant Self-Assembly: A Predictive Molecular Dynamics Approach*, Langmuir, 7 (1991), pp. 2934-2969.

[96] M. Nomura, *Assembly of Bacterial Ribosomes*, Science, 179 (1973), pp. 864-873.

[97] S. Park, J.H. Lim, S.W. Chung, and C.A. Mirkin, *Self-Assembly of Mesoscopic Metal-Polymer Amphiphiles*, Science, 303 (2004), pp. 348-351.

[98] V.N. Paunov. P.A. Kralchevsky, N.D. Denkov, and K. Nagayama, *Lateral Capillary Forces between Floating Submillimeter Particles*, J. Coll. and Interface Sci. 157 (1993), pp. 100-112.

[99] J.A. Pelesko and D. Bernstein, *Modeling MEMS and NEMS*, (2002), Chapman and Hall/CRC Press, Boca Raton.

[100] L.S. Penrose and R. Penrose, *A Self-Reproducing Analogue*, Nature, 1183 (1957), pp. 4571.

[101] H.A. Pohl, *Some Effects of Nonuniform Fields on Dielectrics*, J. App. Phys., 29 (1958), pp. 1182-1188.

[102] J.H. Reif, S. Sahu, and P. Yin, *Complexity of Graph Self-Assmbly in Accretive Systems and Self-Destructible Systems*, Eleventh International Meeting on DNA Based Computers (2005).

[103] I.H. Riedel, K. Kruse, and J. Howard, *A Self-Organized Vortex Array of Hydrodynamically Entrained Sperm Cells*, Science, 309 (2005), pp. 300-303.

[104] R.E. Rosensweig, *Ferrohydrodynamics*, (1997), Dover, New York.

[105] A. Rosler, G.W.M. Vandermeulen, and H.A. Klok, *Advanced Drug Delivery Devices via Self-Assembly of Amphiphilic Block Copolymers*, Adv. Drug Delivery Rev., 53 (2001), pp. 95-108.

[106] P.W.K. Rothemund, *Using Lateral Capillary Forces to Compute by Self-Assembly*, PNAS, 97 (2000), pp. 984-989.

[107] P.W.K. Rothemund and E. Winfree, *The Program-Size Complexity of Self-Assembled Squares*, Symposium on Theory of Computing (STOC 2000), Portland, OR, (2000).

[108] P.W.K. Rothemund, *Theory and Experiments in Algorithmic Self-Assembly*, Ph.D. Thesis, University of Southern California, (2001).

[109] P.W.K. Rothemund, *Algorithmic Self-Assembly of DNA Sierpinski Triangles*, PLoS Biology, 2 (2004), pp. 2041-2053.

[110] P.W.K. Rothemund, *Folding DNA to Create Nanoscale Shapes and Patterns*, PNAS, 440 (2006), pp. 297-302.

[111] P.W.K. Rothemund, *Scaffolded DNA Origami: From Generalized Multi-Crossovers to Polygonal Networks*, Nanotechnology: Science and Computation, (2006), pp. 3-21.

[112] K. Saitou and M. Jakiela, *Automated Optimal Design of Mechanical Conformational Switches*, Artificial Life, 2 (1995), pp. 129-156.

[113] K. Saitou and M. Jakiela, *On Classes of One-Dimensional Self-Assembling Automata*, Complex Systems, 10 (1996), pp. 391-416.

[114] K. Saitou, *Conformational Switching in Self-Assembling Mechanical Systems*, IEEE Trans. on Robotics and Automation, 15 (1999), pp. 510-520.

[115] L.M. Sander, *Diffusion-Limited Aggregation: A Kinetic Critical Phenomenon?*, Cont. Phys., 41 (2000), pp. 203-218.

[116] M.V. Sapozhnikov, Y.V. Tolmachev, I.S. Aranson, and W.K. Kwok, *Dynamic Self-Assembly and Patterns in Electrostatically Driven Granular Media*, Phys. Rev. Lett. 90 (2003), 114301.

[117] N. Seeman, *Nanotechnology and the Double Helix*, Sci. Am., 290 (2004), pp. 64-75.

[118] R.F. Service, *How Far Can We Push Chemical Self-Assembly?*, Science, 309 (2005), pp. 95.

[119] E. Shakhnovich, G. Farztdinov, A.M. Gutin, and M. Karplus, *Protein Folding Bottlenecks: A Lattice Monte Carlo Simulation*, Phys. Rev. Lett., 67 (1991), pp. 1665-1668.

[120] E. Shakhnovich, *Theoretical Studies of Protein-Folding Thermodynamics and Kinetics*, Current Opinion in Structural Biology, 7 (1997), pp. 29-40.

[121] K. Sharpe, *Mathematical Models of Self-Assembly*, Thesis, Dept. of Math. Sci., University of Delaware, (2006).

[122] W.M. Shih, J.D. Quispe, and G.F. Joyce, *A 1.7-Kiolbase Single-Stranded DNA That Fold into a Nanoscale Octahedron*, Nature, 427 (2004), pp. 618-621.

[123] Y.S. Smetanich, Y.B. Kazanovich, and V.V. Kornilov, *A Combinatorial Approach to the Problem of Self-Assembly*, Discrete Applied Mathematics, 57 (1995), pp. 45-65.

[124] B. Smit, K. Esselink, P.A.J. Hilbers, N.M. van Os, L.A.M. Rupert, and I. Szleifer, *Computer Simulations of Surfactant Self-Assembly*, Langmuir, 9 (1993), pp. 9-11.

[125] L.M. Smith, *The Manifold Faces of DNA*, Nature, 440 (2006), pp. 283-284.

[126] Y. Snir and R.D. Kamien, *Entropically Driven Helix Formation*, Science, 307 (2005), pp. 1067.

[127] M. Sperl, A. Chang, N. Weber, and A. Hubler, *Hebbian Learning in the Agglomeration of Conducting Particles*, Phys. Rev. E, 59 (1999), pp. 3165-3168.

[128] J. Stambaugh, D.P. Lathrop, E. Ott, and W. Losert, *Pattern Formation in a Monolayer of Magnetic Spheres*, Phys. Rev. E, 68 (2003), 026207.

[129] J. Stambaugh, Z. Smith, E. Ott, and W. Losert, *Segregation in a Monolayer of Magnetic Spheres*, Phys. Rev. E, 70 (2004), 031304.

[130] R.A. Syms, E.M. Yeatman, V.M. Bright, and G.M. Whitesides, *Surface Tension-Powered Self-Assembly of Microstructures-The State-of-the-Art*, J. Microelectromech. Sys. 12 (2003), pp. 387-417.

[131] J. Tien, A. Terfort, and G.M. Whitesides, *Microfabrication through Electrostatic Self-Assembly*, Langmuir, 13 (1997), pp. 5349-5355.

[132] A. Troisi, V. Wong, and M.A. Ratner, *An Agent-Based Approach for Modelling Molecular Self-Organization*, PNAS, 102 (2005), pp. 255-260.

[133] D. Vella and L. Mahadevan, *The "Cheerios effect,"* Am. J. Phys. 73 (2005), pp. 817-825.

[134] T. Vicsek, *Pattern Formation in Diffusion-Limited Aggregation*, Phys. Rev. Lett., 53 (1984), pp. 2281-2284.

[135] G.A. Voth, B. Bigger, M.R. Buckley, W. Losert, M.P. Brenner, H.A. Stone, and J.P. Gollub, *Ordered Clusters and Dynamical States of Particles in a Vibrated Fluid*, Phys. Rev. Lett. 88 (2002), 234301.

[136] H. Wang, *Proving Theorems by Pattern Recognition*, Bell System Technical Journal, 40 (1961), pp. 1-42.

[137] W. Wen and K. Lu, *A New Net-Like Structure Formed by a Metal/Oil Electrorheological Fluid*, Phys. Fluids, 8 (1996), pp. 2789-2791.

[138] W. Wen and K. Lu, *Pattern Transitions Induced by the Surface Properties of Suspended Microspheres in Electrorheological Fluid*, Phys. Fluids, 9 (1997), pp. 1826-1829.

[139] G.M. Whitesides and B. Grzybowski, *Self-Assembly at All Scales*, Science, 295 (2002), pp. 2418-2421.

[140] E. Winfree, *Algorithmic Self-Assembly of DNA*, Ph.D. Thesis, California Institute of Technology, (1998).

[141] E. Winfree, F. Liu, L.A. Wenzier, and N.C. Seeman, *Design and Self-Assembly of Two-Dimensional DNA Crystals*, Nature, 394 (1998), pp. 539-543.

[142] E. Winfree, *Algorithmic Self-Assembly of DNA: Theoretical Motivations and 2D Assembly Experiments*, J. Biomolecular Structure and Dynamics, 11 (2000), pp. 263-270.

[143] T.A. Witten Jr. and L.M. Sander, *Diffusion-Limited Aggregation, a Kinetic Critical Phenomenon*, Phys. Rev. Lett., 47 (1981), pp. 1400-1403.

[144] T.A. Witten Jr. and L.M. Sander, *Diffusion-Limited Aggregation*, Phys. Rev. B, 27 (1984), pp. 5686-5697.

[145] H. Yan, S.H. Park, G. Finkelstein, J.H. Reif, and T.H. LaBean, *DNA-Templated Self-Assembly of Protein Arrays and Highly Conductive Nanowires*, Science, 301 (2003), pp. 1882-1884.

[146] H. Yan, T.H. LaBean, L. Feng, and J.H. Reif, *Directed Nucleation Assembly of DNA Tile Complexes for Barcode - Patterned Lattices*, PNAS, 100 (2003), pp. 8103-8108.

[147] Y. Zhang and N.C. Seeman, *Construction of a DNA-Truncated Octahedron*, J. Am. Chem. Soc., 116 (1994), pp. 1661-1669.

[148] Z. Zhang and S.C. Glotzer, *Self-Assembly of Patch Particles*, Nano Letters, 4 (2004), pp. 1407-1413.

Appendix A

The Calculus of Variations

Throughout this text, we've seen how energy minimization principles play a key role in understanding self-assembly. Often, the energy for a system can be expressed in terms of an *energy integral*. In this case, several techniques are available for finding the extremals. One of the most important is the indirect method wherein the Euler-Lagrange equation is used to essentially turn the problem into a problem about differential equations. Here, we give an informal derivation of the Euler-Lagrange equation for integrals of the type:

$$I = \int_{x_1}^{x_2} F(x, y, y') \, dx. \tag{A.1}$$

The basic idea is to turn the problem of minimizing the integral over a set of functions into the problem of minimizing a function of a single real variable. Performing a minimization of a function of a single real variable is then a simple calculus exercise. To accomplish this, we first imagine that we know the actual minimizer $y(x)$. Assume that $y(x_1) = y_1$ and $y(x_2) = y_2$. Consider the function

$$y(x) + \epsilon \eta(x), \tag{A.2}$$

where we impose the conditions $\eta(x_1) = \eta(x_2) = 0$. This function is ϵ away from the exact solution $y(x)$ and agrees with the exact solution at the boundary points x_1 and x_2. See Figure A.1. Plug this function into the integral to be minimized to obtain

$$I(\epsilon) = \int_{x_1}^{x_2} F(x, y + \epsilon \eta, y' + \epsilon \eta') \, dx. \tag{A.3}$$

Now, notice that $I(\epsilon)$ is a function of the single real variable ϵ and that $I(0)$ must be a minimum since $y(x)$ was assumed to be the actual minimizer for I! But, $I(\epsilon)$ viewed as a function of the real variable ϵ having a minimum at $\epsilon = 0$ implies that

$$\frac{dI}{d\epsilon}(0) = 0. \tag{A.4}$$

So, let us compute the derivative of equation (A.3) with respect to epsilon, evaluate at $\epsilon = 0$ and set the result equal to zero as required by equation (A.4). We note that

$$\frac{\partial F}{\partial \epsilon} \bigg|_{\epsilon=0} = \frac{\partial F}{\partial y} \eta + \frac{\partial F}{\partial y'} \eta'. \tag{A.5}$$

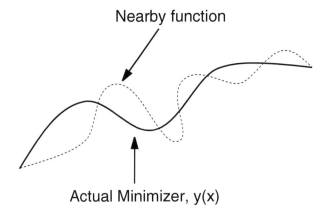

Nearby function

Actual Minimizer, y(x)

FIGURE A.1: A variation from the actual minimizer of the energy. The minimizer, $y(x)$, and the variation, $y(x) + \epsilon\eta(x)$, agree at the endpoints.

hence

$$I'(0) = \int_{x_1}^{x_2} \left(\frac{\partial F}{\partial y}\eta + \frac{\partial F}{\partial y'}\eta' \right) dx \qquad (A.6)$$

and equation (A.4) requires

$$\int_{x_1}^{x_2} \left(\frac{\partial F}{\partial y}\eta + \frac{\partial F}{\partial y'}\eta' \right) dx = 0. \qquad (A.7)$$

Now, integrate the second term in the integral by parts one time to obtain

$$\int_{x_1}^{x_2} \frac{\partial F}{\partial y}\eta\, dx + \frac{\partial F}{\partial y'}\eta \Big|_{x_1}^{x_2} - \int_{x_1}^{x_2} \frac{d}{dx}\frac{\partial F}{\partial y'}\eta\, dx = 0. \qquad (A.8)$$

But, η vanishes at x_1 and x_2 and hence this reduces to

$$\int_{x_1}^{x_2} \frac{\partial F}{\partial y}\eta\, dx - \int_{x_1}^{x_2} \frac{d}{dx}\frac{\partial F}{\partial y'}\eta\, dx = 0 \qquad (A.9)$$

or

$$\int_{x_1}^{x_2} \left(\frac{\partial F}{\partial y} - \frac{d}{dx}\frac{\partial F}{\partial y'} \right) \eta\, dx = 0. \qquad (A.10)$$

Now, this must be true for any η satisfying $\eta(x_1) = 0$ and $\eta(x_2) = 0$. Hence the expression in the integrand multiplying η must be identically zero. That is,

$$\frac{\partial F}{\partial y} - \frac{d}{dx}\frac{\partial F}{\partial y'} = 0. \qquad (A.11)$$

Equation (A.11) is known as the Euler-Lagrange equation for I. Notice that this is an ordinary differential equation for the unknown function y. That is, we have turned the problem of minimizing I over a set of functions into the more familiar and tractable problem of solving an ordinary differential equation defining the minimizer.

Appendix B

Useful Web Sites

In this book we have mentioned several web sites. These are compiled here for easy reference and organized by chapter. Links to all of these sites, plus many more, may be found on the web page for this book:

www.pelesko.com

Chapter 1

The web page below contains links to many of Feynman's lectures.

www.vega.org.uk/video/subseries/8

Feynman's famous 1959 talk that launched the field of nanotechnology is available here:

www.zyvex.com/nanotech/feynman.html

Chapter 2

The web page at the University of Cambridge contains excellent video of bubble raft motion.

www.msm.cam.ac.uk/doitpoms/tlplib/dislocations/index.php

Chapter 3

Nitinol wire is available from many sources. It costs roughly twenty dollars for a five foot roll. This should be sufficient for the protein folding experiment.

www.smallparts.com

www.teachersource.com

Chapter 5

MIT's Star Logo software may be downloaded from:

education.mit.edu/starlogo/

The powerful JavaView Unfolder may be found at:

www.javaview.de

Chapter 6

Video of Saul Griffith's self-folding system may be found here:

alumni.media.mit.edu/~saul/PhD/index.html

The University of Wisconsin hosts a web site with many excellent videos of self-assembling systems.

mresc.wisc.edu

Chapter 7

Video of Swinney's experiments may be found at:

chaos.utexas.edu/swinney.html

Video of Klavins' robots may be found at:

faculty.washington.edu/klavins/

Video of the vibrating particles system may be found at:

www.haverford.edu/physics-astro/gollub/clustering

Chapter 8

A great way to get a feel for DNA is to build your own model. Several kits are available here:

www.thednastore.com/dnastuff/dnamodels.html

Chapter 10

If you want to learn more about the Fibonacci numbers, here is a good place to start:

www.mscs.dal.ca/Fibonacci/

A good resource for all things geometrical is:

www.ics.uci.edu/~eppstein/junkyard/

Appendix C

Glossary

The interdisciplinary nature of self-assembly requires that the vocabulary used in this text be drawn from a variety of disparate fields. To ease the burden on the reader more conversant in mathematics than in biology, or vice-versa, we have compiled this brief glossary.

AFM Acronym for the *atomic force microscope*. This is one of the basic tools of the nanotechnologist. The AFM is able to measure *picoscale* features of surfaces.

Alternating Copolymers This is a polymer, composed of several different types of monomers, that alternate along the polymer chain.

Amino Acid These are the building blocks of proteins. There are approximately twenty different amino acids used in biological systems.

Amphiphile A particle with amphiphilic properties.

Amphiphilic A structure having both hydrophillic (water-loving) and hydrophobic (water-hating) properties.

Base Pairing The complementary pairing of nucleotides in DNA and RNA. In DNA, A binds to T and C binds to G.

Biomimetic The engineering practice of constructing devices that mimic biological systems. The focus is not on creating artificial versions of life forms, but on using the principles nature uses in biology as tools in engineering.

Branched Polymers These are polymers consisting of one main structural branch and small side branches attached to the main backbone.

Calculus of Variations A branch of mathematics that focuses on finding extremals of functionals.

Capillary Bond Refers to the bond between particles created by capillary forces.

Capillary Forces Forces between particles at the surface of a fluid that arise due to the curvature of a meniscus. These forces my be attractive or repulsive.

Cheerio's Effect This is the tendency of small particles floating on the surface of a liquid to be attracted to one another. The breakfast cereal Cheerios, when floating on milk, illustrates this effect.

Coarsening The phenomenon where tiny particles, usually in a fluid, begin to cluster together and form large structures. The original random dispersal becomes *coarser* as time progresses.

Conformation Usually used to describe the particular shape assumed by a protein. In self-assembly, *conformation* refers to the shape or state assumed by any structured particle in a self-assembling system.

Constant Mean Curvature Surface Surfaces satisfying $H = 1$ where H is the mean curvature operator.

Contact Angle A liquid makes contact with a solid surface at a fixed angle. This angle is known as the *contact angle*.

Copolymers These are polymers consisting of more than one type of monomer subunit.

Coulomb Energy The energy of interaction between charged particles.

Coulomb Force The force between charged particles. Also called the electrostatic force.

Critical Micelle Concentration This is the concentration of lipids in a solution where the system undergoes a phase transition and micelles begin to form.

Crystal A solid material with a regular atomic structure.

Cross-Linked Polymers Polymers where individual chains are linked together by side branches.

Degree of Polymerization This refers to the number of monomer subunits comprising a polymer chain.

Dielectrophoresis (DEP) The phenomenon where a dielectric particle in an electric field feels a force due to the rearrangement of charges within the particle.

Diffusion Limited Aggregation A model of crystal growth based on the idea of a random walk. In DLA, a seed particle is placed at the center of a computational grid and additional particles aggregate when they randomly collide with the growing crystal.

DNA Acronym for *deoxyribonucleic acid*. DNA carries the genetic information needed by living organisms. It also serves as a structural material for nanoscience.

DNA Branched Junction The structure formed when a the normally occurring double helix structure of DNA partially unwinds.

DNA DX-Molecule Abbreviation for the DNA double crossover molecule. In this structure a pair of double helices lie side by side and are joined by crossover bonds.

DNA Hairpin Loop The structure formed by a strand of DNA with a sequence that forms its own complement in places. The strand binds to itself leaving an unbound loop at one end. The loop resembles a hairpin.

DNA Sticky End The structure that occurs when one strand of a DNA double helix juts out past the end of the other strand. This leaves a short strand where the molecule may bind to other strands of DNA with complementary base pairs.

Dynamic Self-Assembly As used in this text, *dynamic self-assembly* refers to that subclass of self-assembling processes that lead to stable non-equilibrium structures.

Electrorheological Fluid A fluid that changes its behavior in the presence of an electric field.

Euler's Rule The theorem that say that $V - E + F = 2$ for polyhedra. Here, V is the number of vertices, E is the number of edges, and F is the number of faces.

Ferrofluid A fluid that changes its properties in the presence of a magnetic field. Typically, the fluid is a suspension containing small ferromagnetic particles. When a magnetic field is applied, these particles align along field lines, forming a mesh and giving the fluid rigidity.

Fibonacci Sequence The integer sequence 1,1,2,3,5,8,13 . . .

Folding Funnel A structure in the energy landscape of a protein. The folding funnel has local minima that are easily escapable. This guides the protein to its final folded state.

Folding Pathway A structure in the energy landscape of a protein. A folding pathway specifies the steps taken by a protein as it approaches its final conformation.

Functional This is a map that takes a function as its input and returns a real number. In self-assembly, the *energy functional*, mapping the configuration of a system to the energy of the system, plays a key role.

Geometric Dissection Problem The question that asks: Given a shape in the plane and a series of cuts through that shape, can the resulting pieces be rearranged to form a second specified shape?

Golden Mean The division of a line into two parts such that the ratio of the length of the smaller part to the larger part is the same as the ratio of the larger part to the whole.

Graph Theory The branch of mathematics dealing with the properties of graphs. A graph is a collection of nodes and edges.

Kolmogorov Complexity This is a measure of the complexity of an object. For bit strings it is defined as the length of the shortest computer program needed to reproduce the string.

Hydrophillic Water loving.

Hydrophobic Water hating.

Laplace-Young Law Relates the drop in pressure across a surface to the curvature of that surface. The Laplace-Young Law says that the drop in pressure is proportional to the mean curvature. The constant of proportionality is known as the surface tension.

Law of Mass-Action The principle in chemical kinetics that says that the rate at which a reaction proceeds is proportional to the product of the concentrations of the species participating in the reaction.

Lipid An organic molecule that is often amphiphilic.

Macromolecule A large molecular structure consisting of many smaller subunits. Polymers, proteins, and DNA are examples.

Mean Curvature A measure of the curvature of a surface. The mean curvature is the average of the principal curvatures at a point.

MEMS Acronym for Microelectromechanical systems. These are typically micron sized systems combing both electrical and mechanical components. See [99] for more information.

Meniscus Effect The phenomenon whereby a liquid interface becomes curved in the presence of a solid wall or particle.

Micelle A structure composed of lipid molecules arranged so as to satisfy the amphiphilic constraints of the individual lipids. Micelles form spontaneously in solution at the correct concentration of lipids.

Minimal Surface Surfaces satisfying the equation $H = 0$ where H is the mean curvature operator.

Monomer The basic unit of a *polymer*.

Monte-Carlo Method A numerical method that makes use of random numbers to simulate systems where randomness plays a role.

mRNA Acronym for messenger RNA.

Nanotechnology The science dealing with the characterization and construction of systems characterized by length scales less than 100nm. A nanometer is 10^{-9} meters.

Newtonian Fluid A fluid where the stress is proportional to the strain. Water is an example of a Newtonian fluid.

Nucleotide The basic structural unit of DNA. It consists of a sugar molecule, a phosphate molecule and one of the four bases, A, G, C, and T.

Packing Parameter A semi-empirical parameter used the characterize the shape of amphiphiles. Self-assembled shapes formed from amphiphiles may be predicted in terms of the packing parameter.

Plateau Problem The problem that asks if every closed curve in space can be spanned by at least one minimal surface.

Polycrystal The usual form of a naturally occurring crystal. A polycrystal is a crystal composed of smaller subunits. Each subunit is a regular crystal.

Polydimethylsiloxane (PDMS) A polymer that is often used in self-assembly tile experiments.

Polymer A *polymer* is a long chain macromolecule composed of small individual subunits called *monomers.*

Polymerase Chain Reaction (PCR) A technique for amplifying DNA. It a standard tool in molecular biology, DNA computing, and DNA based self-assembly.

Polymerization The process by which a polymer is constructed from individual monomers. Often, polymerization requires the presence of a catalyst.

Polypeptide Another name for proteins.

Programmed or Programmable Self-Assembly That subclass of self-assembly processes where the particles of the system carry information about the final desired structure or its function.

Protein A large chain molecule composed of amino acids.

Protein Folding Problem The question that asks: Given a sequence of amino acids along a protein chain can you predict the stable conformation of the protein?

Rheology The study of the deformation and flow of matter.

Ribosome A molecular machine found in every living cell. The ribosome is responsible for protein synthesis.

RNA Acronym for ribonucleic acid. RNA is similar to DNA and plays an important role in protein synthesis.

Unit Cell In a crystal, this is the basic subunit than when translated in space reveals the structure of the entire crystal.

Self-Assembly As used in this text, *self-assembly* refers to the spontaneous formation of organized structures through a stochastic process that involves pre-existing components, is reversible, and can be controlled by proper design of the components, the environment, and the driving force.

Sierpinski Gasket A classic fractal that exhibits self-similar structure.

Spanning Tree In graph theory, a spanning tree of a graph is a tree which includes every vertex of that graph. A tree is a graph in which any two vertices are connected by only one path.

Static Self-Assembly As used in this text, *static self-assembly* refers to that subclass of self-assembly processes that lead to structures in either local or global equilibrium.

Stokes' Law The law that says that at low velocities the drag on a sphere moving through a fluid is proportional to the velocity of the sphere.

STM Acronym for the *scanning tunnelling microscope.* This is a basic tool in nanotechnology. Like the AFM, the STM allows measurements at the nanoscale. It is especially useful in wet, biological environments.

Surface Tension The force generated at the surface of a liquid due to the differing environments to which molecules at the surface are exposed.

Thermodynamic Hypothesis The hypothesis that states that proteins choose the conformation corresponding with their lowest energy state.

tRNA Acronym for transfer RNA.

Turing Machine A model of computation based on a finite state automaton and an infinite memory tape.

Turing Universal A model of computation is called *Turing Universal* if it is equivalent to a Universal Turing Machine.

Universal Turing Machine A Turing Machine that can simulate the behavior of any other Turing Machine.

Wang Tiles A set of tiles with colored edges and rules for placement in the plane. Wang Tiles relate tiling to computation and have been shown to be Turing Universal.

Index